Also published by Portland Press Ltd

Techniques in Apoptosis: A User's Guide
edited by T.G. Cotter and S.J. Martin
1996 ISBN 1 85578 076 3

Membrane Protein Expression Systems: A User's Guide
edited by Gwyn W. Gould
1994 ISBN 1 85578 031 3

Microscopy and Histology for Molecular Biologists

A USER'S GUIDE

Edited by J. A. Kiernan and I. Mason

Published by Portland Press Ltd, 59 Portland Place, London W1B 1QW, U.K.

ISBN 1 85578 141 7

British Library Cataloguing-in-Publication Data
A catalogue record for this book is available from the British Library

Although, at the time of going to press, the information contained in this publication is believed to be correct, neither the author nor the publisher assumes any responsibility for any errors or omissions herein contained. Users of this book are advised that prudent laboratory safety procedures should be followed at all times. Opinions expressed in this book are those of the authors and are not necessarily held by the editors or the publishers.

Typeset by Portland Press Ltd
Printed in Great Britain by Information Press Ltd, Oxford.

Contents

Preface

The purpose of this book is to provide instruction in the practice of light microscopy and the preparation of cells and tissues for microscopic examination. It is frequently necessary, especially in research, to make intelligent modifications to methods. With this in mind, the theoretical backgrounds of the various techniques are explained. The worker at the bench will therefore be able to understand the reasons for such things as the composition of a fixative mixture, the pH of a dye solution, or whether immersion oil must be used with a particular objective lens.

The chapters in the first part of the book are concerned mainly with instrumentation: how to use different types of microscope properly, and how to obtain informative images. The second part is devoted to preparative techniques, including methods for looking at living cells, traditional fixation, sectioning and staining, and histochemical procedures for localizing enzymic activities, antigens and specific nucleotide sequences. Techniques for electron microscopy are outside the scope of this book.

The authors of the chapters were asked to write with molecular biologists in mind as the readers. Many in this group of scientists lack the training and experience with microtechnique that were invariably part of the education of biologists of earlier generations. Technique-based disciplines such as histology and molecular biology have converged to create integrated subjects such as developmental biology and neuroscience. It is now frequently necessary for biochemists to look closely at the cells or tissues with which they work; in this book we attempt to tell them how to do so.

John Kiernan
Ivor Mason
London, Canada and London, U.K.
October 2001

Contributors

R.T. Allison, Dental School, Heath Park, Cardiff CF14 4XY, U.K.

A. Boyde, Hard Tissue Research Unit, Department of Anatomy and Developmental Biology, University College London, Gower Street, London WC1E 6BT, U.K.

W.M. Frederiks, Academic Medical Centre, University of Amsterdam, Department of Cell Biology and Histology, Meibergdreef 15, 1105 AZ Amsterdam, The Netherlands

M. Gait, MRC Centre for Developmental Neurobiology, New Hunt's House, King's College London, Guy's Campus, London SE1 9RT, U.K.

G.L. Greenberg, Edge 3D Imaging, LLC, 441 N. 5th Street, Philadelphia, PA 19123, U.S.A.

C. Irving, MRC Centre for Developmental Neurobiology, New Hunt's House, King's College London, Guy's Campus, London SE1 9RT, U.K.

I. Johnson, Molecular Probes Inc., 4849 Pitchford Avenue, Eugene, OR 97402, U.S.A.

J.A. Kiernan, Department of Anatomy and Cell Biology, The University of Western Ontario, London, ON, Canada, N6A 5C1

I. Mason, MRC Centre for Developmental Neurobiology, New Hunt's House, King's College London, Guy's Campus, London SE1 9RT, U.K.

C.J.F. Van Noorden, Academic Medical Centre, University of Amsterdam, Department of Cell Biology and Histology, Meibergdreef 15, 1105 AZ Amsterdam, The Netherlands

S. Van Noorden, Department of Histopathology, Hammersmith Hospital, Du Cane Road, London W12 0NN, U.K.

R. Wingate, MRC Centre for Developmental Neurobiology, 4th Floor, New Hunt's House, King's College London, London SE1 1UL, U.K.

Abbreviations

ABC	avidin–biotin complex
AM	acetoxymethyl
APAAP	alkaline phosphatase–anti-alkaline phosphatase
APES	3-aminopropyltriethoxysilane
BCIP	5-bromo-4-chloro-indolyl phosphate
BrdU	bromo-deoxyuridine
BSA	bovine serum albumin
CCD	charge-coupled device
CI	colour index
CSLM	confocal scanning laser microscope
2D	two-dimensional
3D	three-dimensional
d.p.i.	dots per inch
d.p.m.	disintegrations per minute
DAB	3,3′-diaminobenzidine tetrahydrochloride
DABCO	1,4-diazobicyclo-(2,2,2)-octane
DEPC	diethylpyrocarbonate
DIC	differential interference contrast
DIG	Digoxigenin
DMSO	dimethyl sulphoxide
DTT	dithiothreitol
EM	electron microscopy
EthD-1	ethidium homodimer-1
FITC	fluorescein isothiocyanate
HBSS	Hanks' buffered salt solution
H&E	haematoxylin and eosin
HEMA	2-hydroxyethyl methacrylate
Ig	immunoglobulin
JC-1	5,5′,6,6′-tetrachloro-1,1′, 3,3′-tetraethylbenzimidazolylcarbocyanine
LUT	look-up table
MNA	4-methoxy-2-naphthylamide
NA	numerical aperture
NBT	Nitro-Blue Tetrazolium
OCT	optimal cutting temperature
PAP	peroxidase–anti-peroxidase
PBS	phosphate-buffered saline
PMA	phosphomolybdic acid
PMT	photomultiplier tube
PTA	phosphotungstic acid
SLR	single-lens reflex
TBS	Tris-buffered saline
TdT	terminal deoxynucleotide transferase
TESPA	3-aminopropyltriethoxysilane
TNBT	Tetranitro Blue Tetrazolium
TSM	tandem scanning confocal microscope
TUNEL	terminal transferase-mediated dUTP–biotin nick end labelling

List of methods

CHAPTER 9

CHAPTER 10

CHAPTER 11

CHAPTER 12

CHAPTER 13

CHAPTER 14

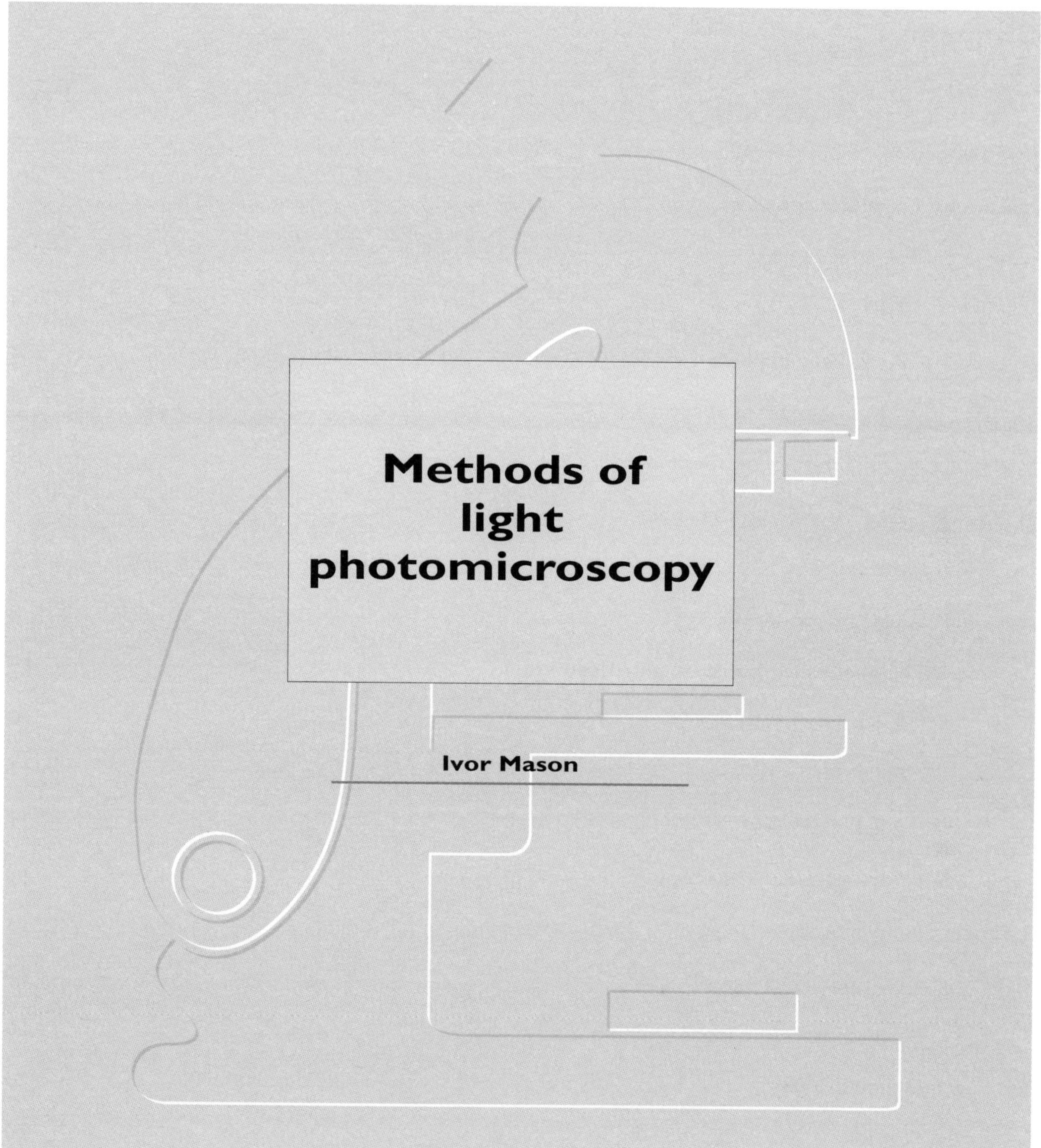

Methods of light photomicroscopy

Ivor Mason

Chapter 1

Page

1. Introduction

Anyone undertaking the techniques described later in this book will, ultimately, need to use a microscope both to analyse their data and to record it photographically. The ability to produce convincing photomicrographs is often crucial in getting results accepted for publication. Indeed, it is clear to me that a manuscript accompanied by clear and attractively presented photomicrographs that convincingly demonstrate the authors' assertions stands the very best chance of being accepted for publication. Therefore, it is essential that the correct use of the microscope is understood if one is not to suffer the humiliation of being asked to present data that is of suitable quality for publication (i.e. figures that support the authors' arguments, are in focus and have the correct colour balance). All microscopes differ to some extent in their various methods of use, although certain aspects, in particular adjusting the microscope to produce optimum brightfield illumination (Köhler illumination), are common to all. Therefore, while this chapter can outline many of the general principles, readers should also familiarize themselves with the manufacturer's instruction booklet for their own microscope.

2. Compound microscopes

2.1. Objective lenses

Modern microscopes will, generally, be purchased with a series of objective lenses covering a range of magnifications between 5× and 60×. A typical series might consist of 5×, 10×, 20×, 40× and 60×, the latter being an oil-immersion lens. Lenses with magnifications of 2.5×and 100× may also be added. Objective lenses on modern microscopes, with the exception of 1× lenses, will be parfocal or approximately parfocal;

this allows objective lenses to be switched without any significant alteration to the focus. Parfocality depends on the correct setting of the diopter of the eyepieces when adjusting the microscope for use (see section 2.4).

All lenses have information concerning their type, magnification, numerical aperture, immersion medium, tube length and coverslip thickness engraved on them. The following sections discuss briefly the significance of these to the user.

2.1.1. Types of lens

Various abbreviations are engraved onto objective lenses to provide information about properties, most notably whether or not they are corrected for spherical aberration and to what degree, and correction for wavelength transmission. Table 1 summarizes these terms and their meaning.

An objective lens can only be used on a microscope of appropriate tube length. The correct tube length is marked on the side of the objective. Generally, the tube length is 160 or 170 mm, however, certain systems e.g. Zeiss Axiophot, use infinity optics marked '∞' on the lens.

2.1.2. Magnification

A lens is also marked with its magnification, although the marked value may be 0.3 times greater or lower than the magnification that it actually achieves. This may be of significance when the microscope is linked to a digital imaging system and is used to calculate distances automatically. In such instances the true magnification should be calculated using a graticule slide and the imaging software.

Table 1. Abbreviations that are engraved onto objective lenses

Term	Description
Achromat	An achromatic lens. Spherical aberration is only corrected for a single colour, and chromatic aberration is corrected for two colours. The remaining aberration can be reduced if yellow–green or green light is used to illuminate the specimen. With this type of lens, colour photomicrographs frequently do not match the perceived colours of the specimen due to the incomplete colour correction. Therefore a poor choice for photomicroscopy.
FL	Fluorite or semiachromatic. This lens is corrected for chromatic aberration but spherical aberration may result in curvature of the image field.
Apo	Apochromatic. This type of lens is corrected for red, green and blue spectra. Apo lenses have high numerical apertures which enhance resolution and have good colour transmission properties. Good colour photomicrographs can be achieved.
Plan	Minimal colour correction although corrected for spherical aberration to give a flat visual field. A poor choice for colour photomicrographs.
Plan Apo	Plan apochromatic. These expensive lenses are colour corrected across the visual spectrum and also for spherical aberration. The lens of choice for colour photomicroscopy.
LWD or ELWD	Long working distance or extra-long working distance high power lenses which allow an increased distance between specimen and lens. Typically used in compound microscopes being used for procedures such as microinjection.
Ph	Phase. The lens contains a phase plate for use in phase contrast microscopy.
Fluor/Neofluar	Lenses for UV-stimulated epifluorescence. Lenses that are corrected within the UV and visual spectral range. The only type of lens suitable for analysis and photography of fluorescent specimens.

2.1.3. Numerical aperture (NA)

The NA of a lens is a key determinant of its quality for both microscopy and photomicrography. The higher the NA of a lens, the greater its light-gathering properties and the greater its spatial resolution.

The NA of a lens is the refractive index of the medium between the lens and the specimen (generally air or immersion oil) multiplied by the *sine* of the half angle of the cone of light entering the objective. As the distance of the lens from the specimen increases so this cone becomes narrower and the NA gets smaller.

The depth of field of a lens is inversely proportional to the square of its NA. Therefore, although resolution is lost with decreasing NA, depth of field increases.

2.1.4. Immersion medium

Most objective lenses are designed for use with a single immersion medium, details of which are engraved on the lens (Table 2). The exceptions are those designed to work in air, and these carry no mark.

The effect of the increased refractive indices of the liquid immersion media is to increase the angle of the cone of light collected by the lens, and thereby improve its NA and resolution. Thus, the highest quality high-magnification lenses are of the oil-immersion variety.

Some objectives can be adjusted for use with a range of immersion media. These have a rotating collar, which is turned to the appropriate position for the medium in use. The collar is often calibrated according to refractive index (see Table 2), and its adjustment compensates for the alteration in refractive index.

However, for best optical performance, the density of the immersion fluid should be matched as closely as possible to that of the mounting medium. Obviously, this is not possible when using lenses that work in air, but the refractive index of the mountant should be as low as possible in such circumstances. In contrast, permanent mountants, such as the popular DePeX, have relatively high refractive indexes and are best combined with an oil-immersion lens.

Table 2. Types of immersion media

Label	Immersion medium	Refractive index
None	Air	1
Oil, oel, HI	Immersion oil	1.52
Glycerin, glycerol	Glycerin, glycerol	1.47
WI, WAS	Water	1.33

2.1.5. Coverslip thickness

Most lenses are engraved with a coverslip thickness value of 0.17 mm; this is the thickness of the standard coverslip for light and epifluorescence microcsopy. Some dry objectives come with a collar that can correct for variations in coverslip thickness and which should be adjusted to maximize contrast.

2.2. Light sources

To obtain optimal optical properties from the microscope, the manufacturer's instructions must be followed to centre and focus the sources of both brightfield (usually tungsten, but sometimes a halogen) and ultra-violet illumination. In order to obtain correct Köhler illumination, the image of the filament of the brightfield source must be focused onto the aperture diaphragm of the substage condenser. In contrast, for epifluorescence, the image of the filament of the mercury-vapour lamp is focused directly onto the specimen. It should be noted that the lifetime of the latter type of bulb is greatly reduced if it is switched on and off rapidly. Once lit it should be left on for at least 30 min, and once switched off, it should not be relit for at least 30 min.

Once focused, light sources should remain aligned for weeks or months, and, therefore, only need infrequent adjustment.

It is sometimes necessary to use incident light from a cooled light source (e.g. Schott) to illuminate a specimen, e.g. to view a whole embryo that has been subjected to non-radioactive *in situ* hybridization. The specimen should be mounted under a coverslip (avoids light dispersal effects of a liquid meniscus) and illuminated obliquely from at least two angles.

> ***Note*** *It may be useful to place a piece of matt black paper beneath the specimen to avoid reflections from condenser lenses.*

2.3. Substage condenser

The primary function of the substage condenser is to focus brightfield illumination onto the specimen. To achieve this, it contains a complex of lenses and an aperture diaphragm. It also contains filters, prisms and annuli that are used in contrast enhancement techniques.

Note *Closing the aperture diaphragm (stopping down) can be used to increase the depth of field of an objective lens. In addition, the aperture diaphragm optimizes contrast, depth of field and NA of the condenser lens.*

Method 1. Adjusting the microscope to produce optimum brightfield illumination: Köhler illumination

Correct Köhler illumination optimizes the microscope illumination to the NA of the objective. This basic adjustment must be performed on every occasion that the microscope is used and for each individual lens or user. Fig. 1 illustrates the typical location of the microscope parts involved in this process.

1. Place a thin specimen, e.g. a slide of a 10 μm-thick stained section, on the specimen stage. It is convenient to have one or two slides suitable for this purpose kept near the microscope. Set the eyepiece collar adjustments to the zero position.
2. Roughly focus a low-power objective on the specimen.
3. Swing in the objective of choice and focus on the specimen.

   A low-power objective is focused first, to ensure that, when a longer, high power lens is swung into place, neither the latter nor the specimen is damaged by contact with the coverslip. This is standard good practice throughout a photomicroscopy session.

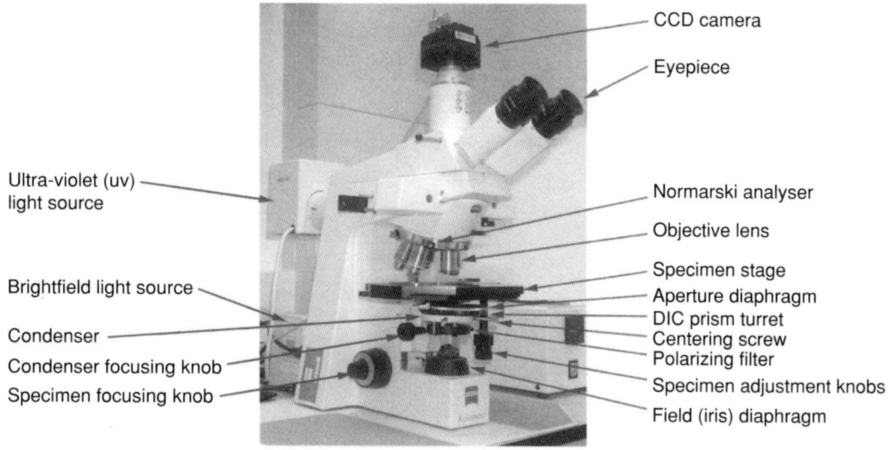

Fig. 1. A typical photomicroscope

The locations of the key elements, whose correct use is essential for good photomicrography, are indicated.

4. Focus each eyepiece individually to obtain optimum resolution. The new position of each adjustment ring can be noted and eyepieces set to this on future occasions.

5. The condenser will frequently carry a range of lenses or differential interference contrast (DIC) prisms. Check that the NA of the condenser lens and the DIC adjustment collar matches the value on the objective.

6. Close the field (or iris) diaphragm to give a small circle of light.

7. Centre this circle in the field of view using the adjustment screws on the condenser.

8. Focus the condenser so that the edges of the image of the field diaphragm are sharp.

9. Open the field diaphragm until it is just visible at the edge of the field of view. Do not open it fully as stray light may enter the light path.

10. The aperture diaphragm is then adjusted to match the NA of the lens as follows. Remove one eyepiece and look down the light tube.

11. Close the aperture (condenser) diaphragm so that 70–80% of the field is illuminated.

Replace the eyepiece and repeat Steps 5–11 every time a new objective lens is selected.

2.4. Techniques for contrast enhancement

All contrast enhancement techniques utilize variations within a specimen, and convert these into variations in light intensity. Examples are light polarizing properties, density and refractive index.

Prior to using any of these, the microscope should have been adjusted for good Köhler illumination (see Method 1).

2.4.1. Normarski/DIC

DIC or Normarski optics detect variations in optical density within a specimen. These local variations are enhanced by prismatic splitting and recombining of polarized light passed through the specimen. Light from the source passes through a polarizing filter (polarizer) and then a prism, both generally located in the condenser. It passes through the specimen and then through a second prism and analyser filter located just behind the objective. It may be necessary to rotate the polarizer slightly in order to maximize the DIC effect.

It should be noted that some biological specimens have polarizing properties themselves. In particular, structures that contain parallel fibres (e.g. skeletal muscle actin–myosin system) are inherently polarizing or birefringent. These structures will appear either very bright or very dark depending on the orientation of their optical axis with respect to the analyser of the microscope.

2.4.2. Phase contrast

Phase contrast optics exploit differences in thickness or refractive index within transparent material, and convert these into differences in light intensity. This technique is only suitable for use with thin specimens. To achieve phase contrast the specimen is illuminated by a ring of light produced by an annulus, the phase annulus, located in the condenser (often at a position denoted by Ph on the condenser turret). The microscope is set up (following the manufacturer's instructions and using the focusing telescope provided) such that the ring of light is aligned with a ring called the 'phase plate' located at the back of the phase objective (all phase objectives are marked Ph). The condenser may contain a number of phase annuli, each matched to the phase plate of a different objective.

2.4.3. Darkfield

Darkfield is useful for observing naturally dark or darkly-stained specimens. It is particularly useful for visualizing silver grains deposited after autoradiography, e.g. during radioactive *in situ* hybridization (see Chapter 13). Darkfield or oblique illumination makes such dark objects appear bright against a dark background. The effect is the result of the condenser directing oblique illumination through a specimen such that the objective lens only receives light if it is reflected or refracted by an opaque object in the field. Transparent regions do not deflect the light to enter the condenser and so appear dark. Some condensers have darkfield optics incorporated into the turret, while on other microscopes, a dedicated darkfield condenser is provided. The procedure for obtaining correct darkfield illumination varies between microscopes, and the manufacturer's instructions should be consulted. However, darkfield can also be obtained in the centre of the image by combining a non-phase objective of 16× or less with a phase contrast ring in the condenser. The objective is focused on the specimen and the condenser is racked down until darkfield is obtained in the centre of the field of view. It is likely that this illumination will not extend to the edge of the field and only the centre is suitable for photography.

> ***Note*** *Dust can be a particular problem when using this approach as it will produce bright specks on the specimen.*

2.5. Fluorescence microscopy

Fluorescence microscopy is a standard analytical approach, most frequently combined with immunohistochemistry, in which multiple fluorochromes can be used to detect several antigens in the same preparation. Most modern microscopes use epifluorescence optics, in which the fluorescent beam is projected through the objective lens which then collects the emitted light from the specimen. Fluorochromes (e.g. fluorescein or rhodamine) that are excited in the ultraviolet or visible range (most fluorochromes are not excited in the ultraviolet spectrum), and emit at a range of wavelengths in the visible spectrum are used. Filter blocks appropriate to the emission spectrum are used to block out other wavelengths, in particular the excitation wavelengths, and it is important to ensure that, if multiple fluorochromes are used, the filter blocks each remove the emission of all but one of them. The analyser filter used for Normarski/DIC should be removed for epifluorescence as it significantly reduces the image intensity.

Autofluorescence can be a particular problem. Certain fixatives, such as glutaraldehyde, fluoresce and should not be used, although treatment with $NaBH_4$ can alleviate the problem to some extent. Some counterstains (e.g. Nissl, Feulgen) should also be avoided for the same reason. The pH of the mountant can also be important; above pH 8.5, fluorescein fluorescence is greatly reduced.

Background fluorescence often increases with time as a specimen is stored and, conversely, emission for fluorochromes becomes reduced. Thus it is important to view and record data as soon as possible. Emission is enhanced by using 1,4-diazobicyclo-(2,2,2)-octane (DABCO) antiquenching mounting-medium, either prepared as below or available commercially.

Materials

- DABCO antiquenching mounting medium
  Dissolve 2.5 g of DABCO (Sigma) in 90 ml of glycerol at 37 °C and dissolve 0.1 g of sodium azide in 10 ml of 1 × PBS. Combine the two solutions, mix well and store at 4 °C in the dark.

3. Dissecting microscopes

Materials

- Sylgard-coated dishes
  - □ Plastic Petri dishes are selected according to the size of the material to be examined. Sylgard resin (Dow Corning) is mixed 9:1 with the supplied hardener, being careful not to introduce air bubbles. The resin is poured into the dish to a depth of 5 mm and left to set

> ***Note*** *Sylgard cannot be used with organic solvents.*

- Agarose-coated Petri dishes
  - □ 1% (w/v) agarose is prepared in 1 × PBS [add sodium azide to 0.1% (w/v) for long term storage, but obviously should not be used when examining living specimens]. Dishes are poured to a depth of 5 mm and stored at 4 °C.

Dissection microscopes have longer working distances and greater depth of field than compound microscopes, but the lower NA of their optical system results in poorer resolution. Much simpler in construction (and use) than compound microscopes, it is nevertheless important to set up a dissecting microscope prior to use in order to both prevent eye strain and to obtain good quality photomicrographs. The dissection microscope generally has much greater depth of field than that of compound microscopes and this allows the eye to compensate more readily if it is incorrectly set up. Thus eye strain is much more commonly encountered when using this type of microscope.

The lack of a condenser system makes the correct adjustment of the optical system much simpler than for compound microscopes, and many modern dissection microscopes have parfocal optics for the various objective lenses. At the start of a session, the microscope should be adjusted as follows.

1. Place a specimen on the specimen stage.
2. Set the eyepiece collar adjustments to the zero position.
3. Roughly focus the low power objective on the specimen.
4. Swing in the highest magnification objective and focus onto the specimen.
5. Focus each eyepiece individually to obtain optimum resolution. The new position of each adjustment ring can be noted and eyepieces set to this on future occasions.
6. On a parfocal system, when the optics have been adjusted for the highest magnification objective, all of the other objectives should be in tolerable focus.

The long working distance and lack of a substage condenser system allow several types of illumination to be used singly or in combination when working with dissection microscopes.

Transmitted light

Light is passed through the specimen; most dissecting microscopes come with such an illumination system integral to their baseplate. This can be used to photograph or examine sections of larger material, e.g. older embryos prepared in the same way as for compound microscopes.

Whole material such as whole embryos may need to be pinned in place and transparent Sylgard-coated dishes (see above) are suitable for this purpose and can be re-used several times.

Incident illumination

Incident illumination is generally provided by an external light source (e.g. Schott), and delivered via a fibre optic system such as a series of 'swan-necks'. This has the advantage that the light does not heat the specimen significantly. When photographing specimens using incident illumination it is useful to have a diffuse or a light-absorbing background. The former can be achieved by placing the specimen in an agarose-coated dish and the latter by having a matt black background. Material can be held in place in such dishes by creating a small trough and sitting the specimen in it.

Sometimes combining both incident and transillumination provides the most satisfactory results. However, care should be taken to avoid shadows when using incident illumination alone; adjusting the angle of the incident light and having two or three sources can achieve this.

Darkfield illumination

A fibre optic accessory for an external light source is now available to produce darkfield illumination suitable for examining and photographing large sectioned material on a dissecting microscope. This is sometimes also useful for examining whole material providing it is not too large.

4. Taking good photomicrographs

This section provides brief notes which detail some of my own experiences in wrestling with the problem of taking good photomicrographs and serves to extend Chapters 2 and 4 by Richard Wingate.

Digital imaging methods are increasingly replacing traditional photographic techniques in microscopy owing to their speed, ease of image manipulation and ability to be imported directly in publication-ready form. The following sections discuss briefly both methods, for which the need to optimize the focus, contrast, resolution and colour of the specimen are the same.

4.1. Basic photomicroscopy techniques

Light from the specimen must be focused on the film of a standard camera or the chip of a digital camera. Most modern compound microscopes divert some of the light from each of the eyepieces to the camera, while dissecting microscopes generally divert all of the light from one of the two eyepieces. It may be possible to check the focus through the eyepiece of a single-lens reflex (SLR) camera or by producing a 'live' image from a digital camera on a screen. Modern photomicroscopes are designed such that camera and eyepieces are parfocal, as long as the diopter of each eyepiece has been correctly adjusted for the observer (see sections 2.4 and 3). Thus it is crucial that the each user of the microscope adjusts the eyepieces for his/her own vision. Failure to undertake this simple step is probably the major reason for the production of poor photographic images. Particular care must be taken when using dissection microscopes, owing to the ability of the eye to accommodate due to their greater depth of field.

Note *For the best photomicrographs, try to use the objective lens with the greatest magnification and numerical aperture.*

Due to the costs of printing colour images, most journals prefer (some insist) that monochrome photographs are provided. Indeed, monochrome has some advantages over colour photography; in particular, monochrome films do not suffer from reciprocity law failure (see below). The 'speed' of a film or its sensitivity to light is indicated by its ASA or DIN value; high values are more sensitive ('faster'), but produce a more grainy negative. Variations in the gamma of the film emulsion is reflected in the contrast of the resulting image. These considerations determine the choice of film for a particular application. Table 3 lists some of the more commonly used monochrome films.

Some workers favour the use of a monochrome film to capture fluorescence images due to its increased sensitivity, and then add back the correct emission colour to produce a pseudocolour print.

One advantage of monochrome imaging is that colour filters can be used to optimize the contrast of regions of interest in a specimen, by removal of colours complementary to that of the filter. A blue–green filter will enhance the brown colour of a DAB (3,3′-diaminobenzidine tetrahydrochloride) reaction product, while a yellow or red filter will enhance the blue colour of a Nitro-Blue Tetrazolium product.

When using colour film, the most common problem is that the colours perceived by the eye do not match those produced in the image. This is

Table 3. Commonly used monochrome films

Monochrome film	Comments
Kodak Technical PAN 125 ASA	Slow, high contrast
Ilford FP4 125 ASA	Slow, intermediate contrast
T-Max 100 ASA	Slow, soft contrast
T-Max 400 ASA	Fast, for immunofluorescence
AGFA Scala 200 ASA	Reversal film for directly producing transparencies

Table 4.	Effect of filters on transmitted light for use with monochrome film			
Filter	**Colour of object of interest in specimen**			
	Red	**Yellow**	**Green**	**Blue**
Yellow	Light	Very light	Light	Very dark
Magenta	Light	Light	Very dark	Light
Cyan	Very dark	Dark	Light	Light
Blue	Very dark	Dark	Dark	Very light
Green	Very dark	Light	Very light	Dark
Red	Very light	Light	Dark	Very dark

because different light sources have different colour temperatures. For example, pictures taken with a standard 'daylight' photographic film will appear reddish when combined with a tungsten light source, as the latter gives off a reddish light. This is not obvious to the observer however. Therefore, it is important to match the colour temperature of the film with that of the light source. Many modern compound microscopes have a 3200 K setting for illumination, which will produce a standardized spectrum of incident illumination that is matched to particular films (e.g. Fujichrome 64T or Kodak Ektachrome 160T).

Daylight films are adjusted for a colour temperature of between 5500 and 6000 K. Some compound microscopes are fitted with a blue filter, which will compensate for this and allow them to be used with a tungsten light source. Daylight films should also be used for colour fluorescence photomicrographs owing to the increased temperature of the UV light source, although monochrome films give better results when only a single fluorochrome is used. Table 5 lists film types used by this laboratory.

A particular problem of colour photography is reciprocity failure, which occurs when exposure times lengthen to more than about 2 s. Under these conditions the effective speed of a film decreases and must be corrected by increasing exposure times accordingly. In addition, there are alterations in the sensitivity of colour pigments in the film, in particular the green pigment. The result of this is a loss of green colour and a purplish tint to the resulting photomicrograph. Although filters can be used to correct this, it is better to avoid such problems by choosing a film of appropriate speed at the outset.

Table 5. Commonly used colour films

Film	Use
Kodak Ektachrome 160T	Standard tungsten illumination
Fuji 64T	Standard tungsten illumination
Kodak Ektachrome Professional	Tungsten illumination with high shutter speeds
Kodak Ektachrome 400 ASA	Fluorescence
Kodak P1600X	Fluorescence

The light meters in cameras generally average the brightness of the whole framed area and select the exposure time accordingly. However, the best photograph is rarely that selected by the meter, normally being either brighter or darker than the averaged light for the field. Therefore, it is advisable to take a range of exposures. Table 6 gives guidelines according to our own experience.

Table 6. Typical exposure adjustments for photography

Method	Technique	Exposure
Transmitted light, e.g. histological slides	Köhler illumination	Overexpose by 1–2 stops (we take 0, 1 and 2 series)
Autoradiographs, e.g. *in situ* hybridization	Darkfield	Underexpose by 1–2 stops (we take 0, −1 and −2 series)
Unstained material, e.g. young embryos	Normarski	Expose according to meter (we take −1, 0 and +1)
Tissue culture cells	Phase contrast	Expose according to meter (we take −1, 0 and +1)
Whole embryo *in situ* hybridization	Köhler illumination	Expose according to meter (we take −1, 0 and +1, but sometimes −2 or +2 is required for very thick or very thin specimens)
Non-radioactive *in situ* hybridization to sections or immunohistochemistry with a chromogenic substrate	Köhler illumination	Overexpose by 1–2 stops (we take 0, 1 and 2 series)
Immunofluorescence	Epifluorescence	Underexpose by 2 stops (we take 0, −1 and −2)
Immunofluorescence: double exposure of same field	Epifluorescence	Underexpose each exposure by 2.5–3 stops

4.2. Digital imaging

For obvious reasons, digital photomicrography is fast becoming the method of choice for routine image collection. The quality of cameras and their accompanying data collection and image analysis/manipulation programs has increased greatly during the past 5 years. Considerations such as the temperature of the incident light remain as described above but are generally corrected by the software after image collection. Digital imaging is dealt with in detail by Richard Wingate in Chapter 2 of this volume.

4.3. Framing

A common mistake is to record data photographically without regard as to how it should be presented in published or seminar form, i.e. to be both aesthetically pleasing and, more importantly, to satisfy a referee. This is a skill that develops with experience. However, if from the outset, the scientist tries to address the issue of what framing and magnification will make the most convincing photomicrograph, they will not go far wrong. Think what you need for a publication before you photograph!

It is worth trying, as far as possible, to fill the frame with the specimen and reduce the area of the photograph that is background. Another useful rule, whether for publication or seminar, is always orientate your specimen in the same way (for example, I always have the anterior end of my embryos facing towards the right). This is particularly useful when giving seminars to those that are either not familiar with the species that you work with or are completely unfamiliar with your field.

I am grateful for the influence of Saville Bradbury during the early stages of my career working with the photomicroscope. Work in the author's laboratory is supported by programme grants from the Medical Research Council and the Human Frontier Science Program, and project grants form the Biotechnology and Biological Sciences Research Council and the Wellcome Trust.

Reference

1. Bradbury, S. (1989) An Introduction to the Optical Microscope. Oxford University Press, Oxford.

Digital and confocal photo-micrography

Richard Wingate

Chapter 2

1. Introduction

A photographic negative may have several hundred discrete brightness levels and several thousand points per inch. Compared with these values, digital photography generally offers lower absolute resolution, but offers several obvious advantages in terms of reduced processing costs, the immediacy of image acquisition, and sensitivity. For most practical purposes, digital resolution is adequate for publication, although most digitally acquired micrographs, printed at 300 d.p.i. (dots per inch), will be unlikely to fill more than a column width of a page. Of the two main forms of digital imaging, video capture has been replaced increasingly by solid-state cameras that use charge-coupled device (CCD) detector arrays.

Scanning confocal microscopy represents a special case in digital image acquisition, in which the particular advantages of laser illumination can be used to generate images of very high optical resolution and unparalleled depth of focus. For a general consideration of digital imaging techniques, see Russ [1].

2. Video cameras

For brightfield- or fluorescence-still photomicrography, video cameras do not provide any advantages over traditional photography or increasingly popular solid-state CCD arrays. However, some observable biochemical changes within cells, such as fluctuations in cytosolic ion concentration, take place on the scale of between seconds and milliseconds. If such events can be recorded in 'real-time', for example through fluorescence labelling, video cameras may be the most effective option for image capture. In video or 'tube type' cameras, the specimen image is focused onto a photosensitive conductive surface forming one end of a vacuum tube. Different light intensities are registered as resistance changes by a scanning electron-beam within the tube and are interpreted as changes in current. The resulting output is 'interlaced' such that each frame consists of two superimposed scans of alternate lines within the image. Video has a relatively low resolution (about 480 lines), but has a high capture rate of 25–30 frames per second. Highly sensitive video cameras can detect very weak fluorescent signals but suffer from a number of characteristic disadvantages. First, 'blooming' is the over-representation of a bright spot in the image owing to the spread of charge through the photosensitive target screen. Secondly, the gamma of the response of a video camera may vary with light levels, progressing, for example, from linear to non-linear as illumination levels decrease. Thirdly, interlace 'tearing' may emerge during playback due to a degradation of the timing signals that regulate alternate scans. It is worth remembering that in 'freeze-frame' mode, most video playback units display only one of the two relevant interlaced images, reducing the *y*-resolution to only 220 lines. Mounting a video camera vertically on a microscope may cause degradation of the system due to a build up of contaminants on the detection target.

3. CCD cameras

Solid-state detectors consist of chips with an array of identical collection units. The most common of these used in photomicrography is the CCD, some of which allow accumulation of photons over long exposures, thereby allowing accurate imaging of very dim specimens. The response of the CCD array to absolute light levels is linear and grey-scale. Colour images can be generated either by sequentially capturing images through three coloured filters (green, red and blue), or by overlaying or interspersing three separate detector arrays. While linear collection properties provide predictable performance, it is sometimes desirable to be able to adjust the linearity of the camera response, allowing for either bright spots (gamma $>$ 1) or low-level noise (gamma $>$ 1) in the image to be suppressed (see below). Image capture consists of two phases. In the first phase photons are collected on an array of detectors (over a duration usually regulated by an external mechanical shutter), while in the second phase, the charge is read off each row sequentially ('progressive' frame capture). Although various designs allow these steps to be accelerated or even simultaneous (for example, by providing a dedicated read-out storage array adjacent or interspersed with the detector array), CCDs are inherently optimized for long exposures and low light levels rather than the rapid sequential image capture of video.

3.1. Factors affecting image collection

3.1.1. CCD design

In choosing a CCD camera several factors have to be taken into account. Resolution (the size of the array), quantum efficiency (the amount of usable light that reaches the detectors) and dynamic range (image depth) are three interacting hardware properties that determine performance. The implications of ever decreasing absolute size of sensing arrays (given as the diagonal length of the chip, e.g. $\frac{2}{3}$ inches or $\frac{1}{3}$ inches), which reduces costs and takes advantages of more compact circuitry, illustrate how these properties interact. Smaller, more closely packed detectors are more sensitive to spread of charge between neighbours, thereby introducing 'blooming' effects. Isolating pixel elements to prevent blooming introduces non-sensing boundaries that reduce quantum efficiency (some commercial chips have an efficiency of only 12%) and, therefore, require greater incident light levels. Reduced pixel size also leads to a lower 'well capacity' (the number of electrons stored in each detector) and, therefore, a lower absolute dynamic range. Both spectral and absolute sensitivity and quantum efficiency can be enhanced by thinning the structure of the chip and illuminating the detector array from behind (backlighting). The spectral sensitivity of most CCDs is not uniform and, in particular, may be biased towards infrared detection. In practice, these issues become limiting factors for image capture at the limits of detection ranges. Experimenters should, however, be aware of the trade-off between costs, speed of image capture, pixel resolution and pixel size.

3.1.2. Noise

CCD arrays also suffer from two main sources of noise, which can lead to image degradation. The ideal chip should contain up to 2 million identical detectors. However, this is rarely the case: rogue pixels may introduce hot- or dark-point defects, or pixel 'traps' where the read-out of a row is attenuated by an irregular element, thereby darkening an entire line within the image. The lower the quality of the chip, the more post-collection software processing of the image is required to correct these hardware errors. Additionally, like any electronic circuit, CCD chips suffer from stochastic background noise originating both at detector level (the 'dark current' produced in the absence of light) and read-out noise in the pre-amplification of CCD output. Both can be reduced or eliminated by reducing the operating temperature of the CCD camera to below −60°C. Read-out noise can also be minimized by reducing read-out rates. Cooled CCD cameras are required where the signal-to-noise ratio (which increases with signal amplification, or gain) becomes critical (such as fluorescence photomicrography) and generally provide improved image quality. Random fluctuations can also be removed by averaging a number of scans of the same field of view (Kalman filtering). With more averages, random variations in the image become smoothed, while real image features are preserved. The improvement of a given signal-to-noise ratio is proportional to the square root of the number of averaged frames. Both hardware design and attendant limitations have implications for the way in which image capture can be optimized.

3.2. Optimizing digital image collection

Just as in conventional film photography, the range of lights levels must be matched to the solid-state camera, so that the dynamic range of the collector (emulsion or pixel) is used optimally. This involves adjusting lighting levels of the specimen and the sensitivity of the detector. Collecting multiple images or background reference images allows digital noise to be reduced and image-independent artefacts to be removed, respectively. While digital images are easy to 'correct' after collection, this usually leads to a loss of some aspect of information, and it is always better to optimize the image at the point of collection.

3.2.1. Setting gain and black level

The well capacity of an individual detector determines the dynamic range of the array, which may be as high as 2^{16} discrete levels, or 16 bits. The relevant information in most microscope specimens will be detected over a fraction of this range and incident light levels must be mapped to the response properties of the array. Superficially, this translates simply into adjusting the sensitivity or gain of the CCD until a subjectively reasonable picture is achieved. However, it is possible to be misled by the properties of computer monitors and the adaptive capacity of the visual system, such that subjective decisions translate into a variable final quality. Fortunately, most image collection software allows for an objective decision on sensitivity adjustment to be made, either by displaying the histogram of incident light intensity, or utilizing a colour look-up table (LUT) to show parts of the images which are clipped (fall outside the detectable range).

Fig. 1 shows an 8-bit black-and-white photomicrograph of the central nervous system of the fruit fly (*Drosophila*), which demonstrates how the quality of an image can be tuned to fit available (2^8) grey shades. A schematic histogram of light intensity shows whether the full dynamic range of the detector is being used, to the left, while the resultant image is shown to the right. In Fig. 1(a), the histogram fills the available grey scales and the image shows a range of detail. In Fig. 1(b), the black level is set too high and the gain is set too low to capture the range of intensities in the image, resulting in clipping. Software will often allow the saturation at either end of the intensity scale to be revealed by applying a colour LUT, which represents 'clipped' areas of the image in contrasting colours. In Fig. 1(c), the image occupies only a fraction of the

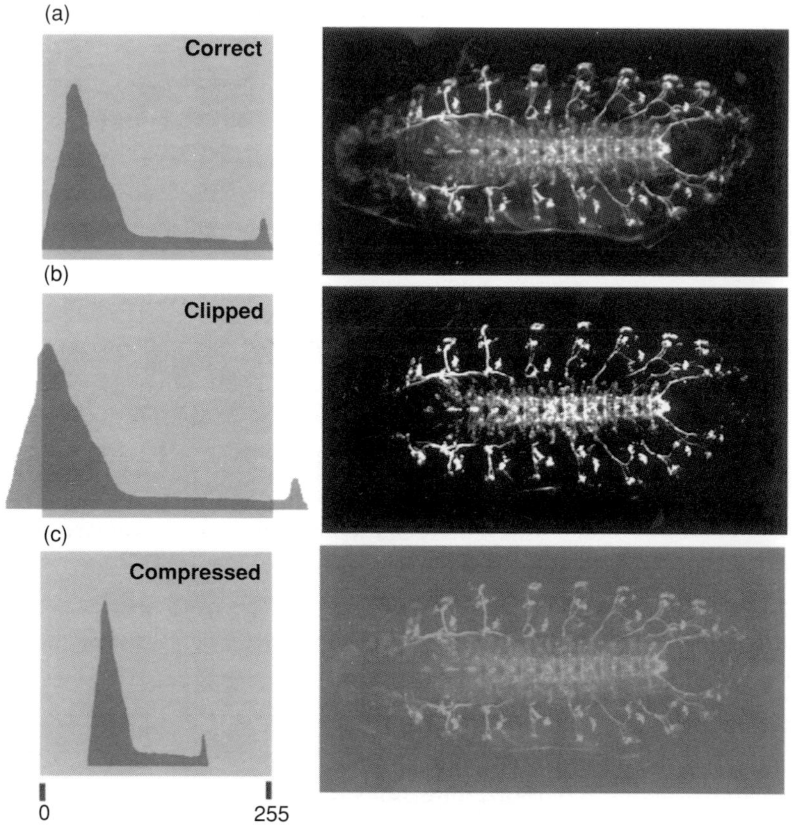

Fig. 1. Examples of correctly judged gain and black level controls in collection of a digital image of the central neurons of fruit fly (*Drosophila*).

Schematic representations of a histogram of intensity distribution are shown on the left, with the corresponding to the micrographs to the right. (a) An image where pixels values fall precisely within the 256 grey shades available in an 8-bit image. (b) The effects of 'clipping', when values fall outside the available ranges of intensities. (c) The washed-out appearance of an image which has not utilized the full range of available tones. The Drosophila embryo was stained with a primary antibody against a neuronal marker and a fluorescent secondary antibody, and was kindly provided by Dr A.M. Myat (MRC Centre for Developmental Neurobiology, King's College London, U.K.). The image was captured on an MRC BioRad 500 confocal microscope.

available intensity levels and has a washed-out appearance. Reducing the black level and increasing the gain will stretch the histogram to fit the available grey scale. A colour LUT is useful in this respect for determining the point at which clipping starts to appear. 'Contrast stretches' can be performed after the image has been acquired; however, this results in no real gain in information and the image should be optimized at the point of collection.

3.2.2. Non-linear gamma adjustment

In some cases, the range of incident light is significantly greater than the dynamic range of the collector. For example, increasing the sensitivity of the array in order to adequately represent a relatively faint part of the image may result in blooming or smearing within a brighter part. Therefore, for some detectors, it is also possible to perform a non-linear expansion or compression of the collection histogram in order to boost or suppress either the high or low ends of the intensity scale (changing the gamma of the detection array). Fig. 2 shows the transformation of detector sensitivity from a linear response (Fig. 2a) and its effects on the final image. To the left, the output of the detector is plotted against the input intensities from the image as an LUT. A positive gamma adjustment (Fig. 2b) enhances fainter parts of the image by expanding the lower-intensity band into a greater number of grey scales. A negative adjustment (Fig. 2c) suppresses low-intensity levels and might be used for suppressing low-intensity noise. In either case, the effect is to increase the representation of part of the image considered to be interesting at the expense of another range of intensities.

3.2.3. Averaging, binning and light levels

As discussed above, averaging successive scans reduces random noise within the detector and results in a cleaner image. If the specimen is poorly illuminated or weakly fluorescent, then the sensitivity of the detection

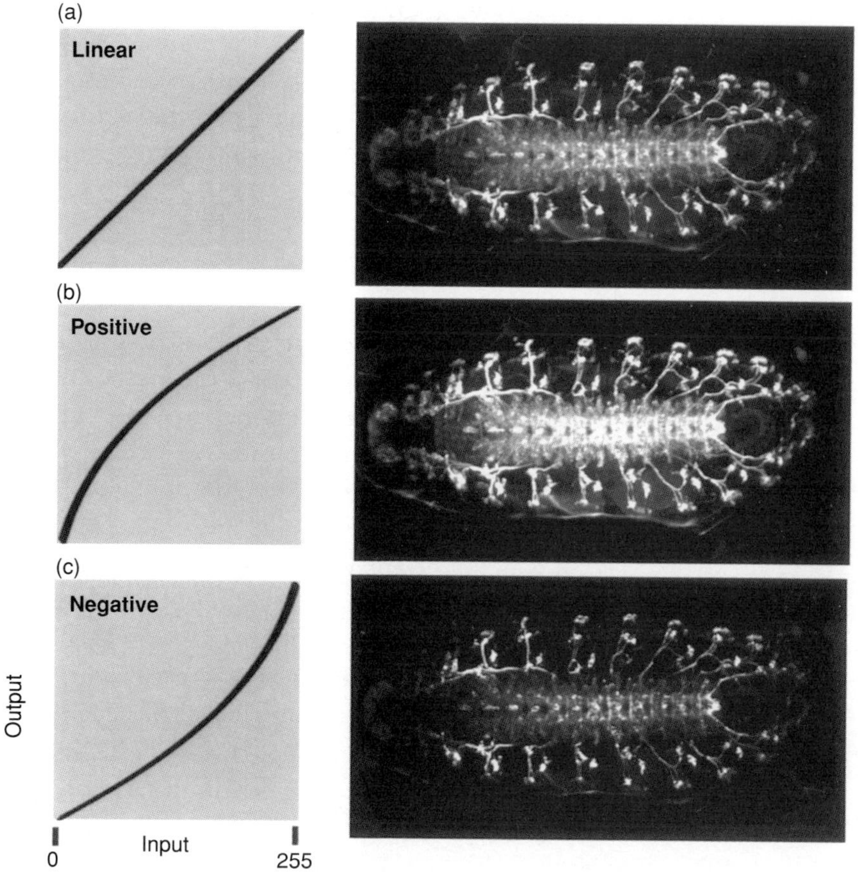

Fig. 2. Examples of how adjusting the linearity (gamma) of a collection device can enhance (b) or suppress (c) fainter objects within the image.

The graphs (on the left) show deviations from a linear collection ramp (top left). Each transformation takes the form of an LUT, in which the output of the detector is plotted against the input spectrum of intensities.

array must be increased, thereby raising the levels of background noise and demanding a higher number of scans to achieve a clean image.

Binning adjacent pixels in blocks of predetermined sizes also provides a means of removing noise by averaging values across the image rather than over time. This results in a lower-resolution image, but in relatively faster capture rates. Binned, low-resolution grey-scale images are often used to provide a rapid stream of representations for positioning and focus. For cheaper CCD arrays that are neither cooled nor capable of averaging, maximizing the incident light levels becomes particularly important.

3.2.4. Colour balance

Colour CCD images are either a composite of the output of three different detector arrays or of the output of the same chip exposed using different filters mounted in a filter wheel. White balance may be determined for each exposure or calibrated in a single measurement. For the latter, it is important to remember that as the voltage across a tungsten filament is altered, the colour of light emitted changes. While it may not be apparent to the observer, changing the voltage of the illuminator will alter the colour of the image. If we consider this, and the signal-to-noise ratios discussed in Section 3.1.2, it seems reasonable to illuminate the specimen with the brightest, constant illumination as for conventional photomicrography ('photo' or '3200/3400' setting on a photomicroscope). Any necessary reductions in intensity can be introduced with neutral-density filtering.

3.2.5. Background image subtraction/division

Static defects within the optical pathway or CCD array, and unevenness of illumination, can be corrected to some extent by recording a reference image that can then be removed digitally from the specimen image. Although, for transmitted brightfield illumination, the condenser lens can be aligned so that a relatively even field can be produced, lenses will always tend to absorb more light at the centre of the image. Given that a digital array has linear collection dynamics, the brightness across the field can be 'levelled' by dividing the specimen image by the reference image. However, this leads to a decrease in the dynamic range of the original data and, therefore, every effort should be made to provide a constant field of illumination for the reference and specimen images.

In certain circumstances, a background reference may be difficult to obtain, in which case various approaches can be taken to level an image ('shading correction'). For example, low-frequency variations in intensity can usually be assumed to be associated with asymmetric illumination, and can be detected and removed by frequency transformation.

3.2.6. Reducing blurring in fluorescent images: deconvolution

When fluorescent labelling is distributed through the depth of an otherwise transparent specimen, only part of the image will be in focus on a conventional optical microscope (see, however, Section 4 on confocal microscopy). To some extent, blur can be addressed by applying an 'unsharp mask', which has a similar effect to a LaPlacian filter (see Section 5.2). An unsharp mask is an out-of-focus negative of the image that is subtracted from the original to bring out local intensity variations. This also reduces the overall contrast of the image. A more sophisticated and computer-intensive approach, available on systems with a motorized focus, is to collect a series of images at different focal depths within the specimen and then 'deconvolve' a given optical plane. Three-dimensional deconvolution relies on being able to estimate the point spread function (the degree of blurring of a perfect point in the image plane) of a given objective lens. The frequency transform of the point spread can then be used to reconstruct a sharp image by division across the frequency spectrum. The degree of blurring for each pixel is calculated locally across successive optical planes, allowing a fairly accurate rendering of the image. Inherent noise in the image causes serious problems for deconvolution algorithms that have to be compensated for by 'Wiener' filtering and an accompanying loss of resolution.

3.2.7. Special considerations with polarized light

Some microscope imaging uses polarized light to highlight particular structures within a specimen, or an assembly of polarizers and prisms to enhance contrast in otherwise transparent material (Nomarski optics; Section 2.4.1). Some digital systems inadvertently introduce polarizing effects that may interfere with optical image enhancement, e.g. a polarizing filter may screen a CCD chip. Although a Nomarski image is apparent through the eye-pieces of the a microscope, it may appear dark on an attached digital camera because an internal filter is orientated at 90° to incoming polarized light from the specimen.

4. Confocal laser-scanning microscopy

By using a scanning laser beam to illuminate a fluorescent specimen and by collecting light through a pin-hole, it is possible obtain optical sections through thick sections or wholemounts, with no out-of-focus blur, producing an image of very high spatial resolution. Stacks of optical sections can be projected digitally to produce an extended focus image through the depth of a specimen using objective lenses that, nevertheless, have a high numerical aperture. Stacks, or volumes, can also be used to create three-dimensional reconstructions of the specimen.

Confocal microscopy relies on only a single point within the specimen being illuminated at any given time. For a conventional microscope, this would be the equivalent of using a condenser with the field diaphragm closed down to a diminishingly small pinhole. The resulting point image is viewed with a perfectly optically matched objective, through an 'exit' pinhole of the same diameter. The three points — the illumination pinhole, the focus of illumination within the specimen, and the exit pinhole — are said to be 'confocal' [2]. Light transmitted from within the specimen, from any point lying outside the precise focus of the illuminating cone of light, will be deflected to a greater or lesser extent. Deflected light will pass through the objective but will miss the exit pinhole and not be observed. Thus a sharp image of a single point can be produced with no scattered light from different focal planes or neighbouring tissue. To reconstruct the field of view within a given optical place, the point of illumination must be scanned across the specimen in a raster-like manner. For the laser-scanning microscope, this is achieved by substituting the condenser pinhole for a high-intensity, monochromatic laser beam. The problem of achieving perfectly matched optics on either side of the specimen is solved by epi-illumination, in which both incident and reflected light are passed through the same lens, i.e. the objective. The resultant grey-scale image is recorded by changes in voltage in a photomultiplier tube (PMT). Automated collection of a number of optical planes, by means of a z-axis focus motor, allows the entire depth of the specimen to be sampled. While this is not the only solution to achieving a coherent confocal image, scanning laser epi-illumination remains the most widespread format for developmental biological research.

4.1. When to use a confocal microscope

Any fluorescent specimen can theoretically benefit from confocal imaging. The observable, minimum resolvable distance is inversely proportional to the size of the illuminated field. By reducing the field to a vanishingly small spot with a scanning laser beam, the spatial resolution of an image is maximized. By optically sectioning fluorescent material, confocal microscopy allows subcellular localization of fluorescent label without physically sectioning material. By projecting a stack of optical planes, a sharp image can be obtained through a thick fluorescent specimen without the use of deconvolution algorithms (see Fig. 3 and Section 4.2.6).

The effectiveness of a confocal microscope relies on the optical precision of its components. At lower magnifications, the theoretical advantages of confocal imaging are decreased because of the lower numerical apertures of objective lenses. However, where the specimen is thick or occluded by superficial, intense labelling, the confocal image is still clearer than with conventional fluorescence. The relatively slow scan speeds and potentially destructive effects of laser illumination usually compromise the confocal imaging of living cells. In most cases, an image is acquired over a number of seconds (the time taken for the laser to raster scan the field of view), or possibly longer depending on the degree of Kalman filtering that is required (see below).

4.2. Collecting a confocal image

The details of collection will vary from system to system. The choice of laser determines the available excitation wavelengths and, therefore, the choice of fluorochromes that can be used. Acousto-optical tuneable filters are used increasingly to select from available excitation wavelengths and set their intensity, replacing conventional excitation and neutral-density filtering. Most systems still employ a recognizable choice of dichroic mirrors and filters to select the range of emitted light. As with other forms of digital imaging, a small set of fundamental principles apply to the operation of confocal microscope PMTs.

4.2.1. Microscope and specimen preparation

Confocal microscopy is dependent on good mechanical alignment and high quality components. It is critical that the relative positioning of the emission and exit pinholes is stable. External factors, such as mechanical vibration, can be minimized by careful positioning of the microscope on an anti-vibration table, and can be compensated by improved design; modern scan-heads can be readily moved between microscopes without alignment adjustment. The optical quality of the microscope is also important for maintaining pinhole registration. In particular, the objective lens, which also functions as condenser, must be of high numerical aperture and well corrected for both colour and flatness of field in order to minimize deflection as different laser wavelengths scan over a wide field of view. From the discussion above, it is clear that any scattering of emitted light will result, not in blurring, but in a loss of signal intensity as information is deflected away from the exit pinhole. Systems with infinity optics minimize internal light scattering within the microscope tube and allow a larger exit pinhole. Finally, light path deviations introduced within the specimen by variations in refractive index (Chapter 1: Section 2.1.4) can be minimized by an appropriate choice of mountants and objective lens immersion medium. At high magnifications, immersion lenses give, approximately, a 1.4-fold better axial resolution than an equivalent dry objective of high numerical aperture.

4.2.2. Laser excitation

For most scanning lasers, different combinations of krypton, argon, neon, helium and helium/cadmium give different ranges of excitation and determine which fluorochrome can be used. Appropriate filtering gives discrete excitation wavelengths and limits the intensity of the light source. In principle, specimen photobleaching is reduced when the lowest sufficient illumination is used and gain is maximized. This also increases the number of scans required for effective Kalman filtering, but repeated illumination at low intensity is less deleterious than exposure to high-intensity laser light.

4.2.3. Setting pinhole size

Reducing exit pinhole size excludes more out-of-focus light and produces a fainter, but sharper, image. This loss of information is compensated by collection of a *z*-series. However, for low-power lenses with a low numerical aperture, the gains in axial resolution obtained by reducing the pinhole to a minimum are redundant because of the lower optical resolution of the objective. Similarly, an 'optimally' adjusted pinhole for a

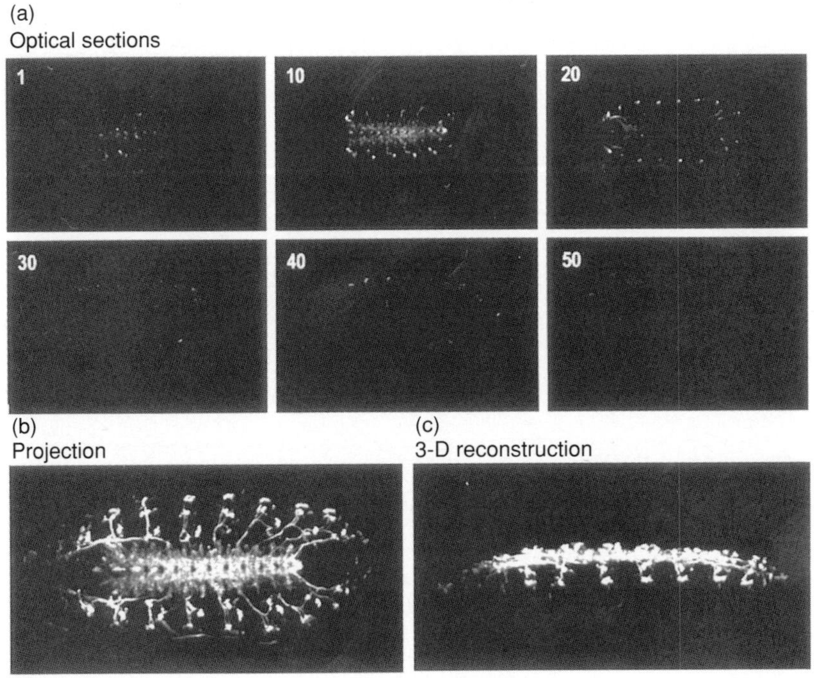

Fig. 3. Composition of a confocal micrograph projection

(a) A selection of optical planes collected at different focal depths within the specimen. A total of 65 planes were used to sample the entire depth of the preparation, and an extended-focus view was constructed by a digital projection of the optical sections (b). Because each stack of confocal images represents a volume of data composed of cubic 'voxels', with appropriate software the image can be rotated as a three-dimensional (3-D) object in virtual space (c).

given objective lens may exclude too much of the useful light, which is inevitably scattered due to the heterogeneous refractive indices of biological material. Determining a theoretical optimum for pinhole size can be achieved by directly observing the point spread of illumination (see Section 4.2.7).

4.2.4. Setting gain and black level

As is the case with other forms of digital image acquisition, the sensitivity of the PMT must be optimized with respect to the emission spectrum using a histogram or colour LUT approach (see Section 3.2.1 and Figs 1 and 2). By optimizing the gain, black level and linearity of the PMT, the full range of available tones can be used to represent the image.

4.2.5. Kalman filtering

Averaging successive scans reduces quantal noise and PMT noise. For a given gain, the number of passes required to minimize random fluctuation in pixel intensity can usually be estimated by monitoring the gradual sharpening of the image until minimal changes are observed.

4.2.6. Collecting a z-series

Stacks of images can be used to produce a two-dimensional projection which gives extended focus through a thick specimen, or can be analysed as a volume of data to produce a three-dimensional reconstruction. For the former, it is self-evident that the entire depth of the optical stack should be represented, but this can be achieved with a relatively broad spacing of optical planes, dependent on the size of the exit pinhole (larger pinholes give greater depth of field but lower resolution). Fig. 3 shows a selection of optical planes which, when projected, give a sharp, in-focus image of the *Drosophila* central nervous system. It is important to note that each individual optical plane is faint. When the pinhole aperture of the detector is closed down, the specimen may disappear from view if the focal plane is not in register with the specimen. For three-dimensional reconstruction, it is usually convenient that the spacing between layers should at least be equal to the *z*-dimension of a theoretical cubic pixel, or 'voxel', because software 'interpolation' of missing planes gives poor results in most biological material.

4.2.7. Optimal resolution: matching pinholes, pixel size and Nyquist sampling

For various reasons, some of which are discussed in Section 4.2.3, biological material presents near-insuperable practical obstacles to achieving the theoretical optimal optical sectioning and spatial resolution of the laser-scanning confocal microscope. Nevertheless, it is worth considering the factors that, theoretically, determine how pixel size, and pixel size with respect to the area covered on the specimen, can be set by empirical criteria. These involve setting the size of the detection pinhole and corresponding pixel array to correspond with the limitations on resolution set by diffraction.

First, pinhole diameter must be matched to the size of the 'Airy disc' associated with spread of a point of illumination at a given wavelength and for a particular numerical aperture. An Airy disc takes the form of a series of light and dark diffraction rings radiating from the centre of the point; two adjacent points can only be discriminated if separated by a distance at least equal to the radius of the most proximal, dark Airy ring. In some systems, the Airy disc for a particular objective can be directly observed using a reflectance filter block to project light onto a screen in the plane of the exit pinhole. Therefore, the pinhole should be set so that it covers half the diameter of the first dark ring within the Airy disc (the minimum discrimination distance). Thus, the size of the optimal pinhole varies empirically with numerical aperture, magnification, and excitation wavelength.

Secondly, the size of the area covered by each pixel should be matched to the Nyquist criterion. Using the zoom factor available on most microscope systems, the pixel array can be mapped optimally to the image if each detector samples an area at least 2.3 times smaller than the resolution limit of the optical system. Therefore, for a given wavelength and objective, the zoom function can be set so that the pixel size is the radius of the closest dark ring on the Airy disc divided by 2.3. By comparison, a lower sampling rate (equal to the radius of the dark ring) is optimal for the *z*-axis. Paradoxically, this requires that the zoom factor is often greater for a low-magnification objective. A tabulated set of zoom values for typical objectives has been provided by Centonze and Pawley [3].

5. Digital image manipulation

Image processing software gives great scope for altering micrographic images, but it cannot substitute for good photographic technique at the time of collection; any manipulation of an image after collection will usually result in a loss of contrast and may introduce false resolution artefacts. As a basic rule, adjustments should be carried out uniformly across the image and be restrained. Selective alteration of image details, particularly pixel cloning and 'cut-and-paste', represents an intrusive alteration of data. If these are used it is suggested that their application and scope be mentioned in any publication (for guidelines see Saper [4]).

A variety of different algorithms are available in most software packages to enhance image features or suppress elements falling outside a defined spatial, intensity or colour threshold.

5.1. Noise reduction (median filter)

Noise that cannot be removed by averaging during image collection can be alleviated by 'neighbourhood' operations. These analyse values from a block of pixels to normalize the pixel at the centre of the block. The size of the block (described as a 'kernel' for some functions), and the details of the operation determine the overall effect. Of the various neighbour operations, the median filter, a simple but effective noise reduction filter, replaces the intensity value of a given pixel with the median value of its neighbours. When applied across an image this effectively removes rogue pixels of high or low value ('shot noise'). The median filter is the preferred method of *post hoc* noise reduction, as it preserves boundary positions and brightness differences that can be lost in other neighbourhood 'smoothing' functions (averaging filters and Gaussian blurring).

5.2. Sharpening

A variety of filters can be used to detect contrast changes within an image and to emphasize boundary features. For example, a simple neighbourhood operator can be used to equalize pixel values across shallow gradients and to boost differences where contrast gradients are steep. The image produced by this kind of 'LaPlacian' operator is subtracted from the original image to create a sharpened version. LaPlacian filters tend to accentuate single pixel variations, such as shot noise, and respond best to sharp discontinuities. Alternative strategies are based on kernels that calculate the first-order differential of the image, e.g. Sobel and Kirsch operators. These are less sensitive to pixel noise, but may exclude real image data if the kernel size is greater than the features of interest within the image.

5.3. Brightness and contrast

A good photomicrograph will not need *post hoc* alteration of brightness and contrast, which inevitably degrades the quality of information within the image. However, there are occasions when the useful information within an image is either compressed to a fraction of the range of available grey scales or is poorly represented. Judging brightness or contrast by eye is risky. Monitors also vary with respect to their own colour balance, with contrast and brightness contributing to error. As with image collection, care must be taken not to saturate pixels within the image. It may be possible to reveal graphically 'clipping' at either end of the brightness scale (e.g. in Adobe Photoshop, as a contrasting overlay colour) so that pixel values are kept within the displayable range. The expansion of data from a small range of grey scales to occupy the full depth of an image cannot increase the amount of information and is, inevitably, a poor replacement for setting the camera correctly in the first instance.

Contrast and colour corrections can also be seen in terms of transformations that map an original set of pixel values to a new set of values. Plotting the original against transformed values generates a curve, an LUT.

5.4. LUTs

The stored brightness and colour value of a given pixel can be converted into a new value according to the transformation given by the curve in an LUT. For example, a linear transfer function with gradient of 1 (identity) maps each pixel value to itself. Any deviation from linearity alters the brightness and contrast of the image. If a grey-scale is mapped to separate red, green and blue brightness curves, it will remain monochromatic while the curves are in register. If the three curves are altered independently of each other a 'pseudocolour' image is produced. Changes in brightness, contrast and colour can therefore be seen in terms of LUT transformations and most software packages allow user-defined LUTs to be defined and stored so that a range of images can be treated identically.

Colour LUTs give an opportunity to correct chromatic variations that might emerge through non-linearity in camera performance. As above, altering colour is highly subjective and runs of the risk of pushing pixel intensity beyond displayable thresholds or colours out of the printable gamut (see below).

Colour LUTs also allow pseudocolour to be applied to black-and-white images and, therefore, are widely used in fluorescent image manipulation.

Simply colouring a monochrome micrograph to simulate an image taken in colour has negligible benefit and degrades image quality. However, pseudocolour is extremely useful for identifying intensity bands within an image or distinguishing two merged images of the same field of view, e.g. when contrasting multiple fluorochromes or structures at different focal planes (see Wingate [5]).

6. File formats for digital images

The rapidly falling cost of image storage and, in particular, the introduction of the CD burner, have reduced the need to store digital images in a format that minimizes file size. Reductions in file size are achieved characteristically by either reducing image depth (the number of intensity levels) or by digital compression. An 8-bit image depth (255 grey shades) provides reasonable quality, although a 12-bit image is the accepted minimum for accurate measurements of optical density. If image data are stored in a 'raw' file-format, image depth and pixel resolution will be the only factors determining size. The addition of colour information generates a corresponding increase in file size. However, compression of file size is achievable without loss of data. For any given image, a number of feature elements will be repeated numerous times, and the data can be described in shorthand simply by listing the occurrence of these redundancies. Different compression algorithms are designed for different image compositions and may be less or more successful with micrographs. However, the highest compression ratios are achieved when some of the data are discarded as they are transformed ('lossy' compression).

Lossless compression formats (e.g. GIF, PICT and TIFF) use run length encoding to identify sequences of identical pixels, and replace them with a single description of the pixel and details of its occurrences. This kind of compression works well with computer-generated illustrations, but has very little impact on an image with shading and colour gradients. The highest compression is thus achieved with binary images.

TIFF designates a variety of different data arrangements with a variable degree of compression.

Lossy compression formats (e.g. JPEG) achieve much higher compression ratios with photomicrographs by analysing frequency distributions within the image. The format is slightly kinder to shading values than colour, in accordance with the emphasis that human vision places on different aspects of an image. The compression/decompression algorithm preserves gradients of colour and intensity but tends to degrade sharp edges, corners and lines (where high-frequency description is required). Degradation at these areas involves loss of contrast and detail as well as the introduction of false colours. Because this type of compression initially subdivides the image into blocks of pixels, decompression sometimes introduces 'quilting' artefacts into the image. Lossy compression formats are, therefore, not suitable for the storage of primary data but are ideal where a typical photomicrograph needs to be stored or transmitted as a small file (thumbnails, e-mail attachments, web-pages etc.).

7. Printing computer images

While many journals accept images in a digital format, a printed image is still required for the review process prior to publication. More importantly, review figure copies may be used to balance tone and/or colour in the commercial printing process that generates the final copy. The printed image is always of a lower quality (resolution and image depth) and, if in colour, of poorer chromatic range than the digital image.

Printing quality and resolution, in terms of d.p.i., are often viewed as synonymous, and journals may specify a minimum d.p.i. for submitted images. Although a d.p.i. value is proportional to resolution on screen, it can be misleading in print due to the different mechanisms by which tonal and chromatic variation are reproduced. For example, half-tone black-and-white images rely on varying the density of ink to produce grey shades. In laser and ink jet printers, groups of dots are grouped together in half-tone cells, each of which defines the tone at a particular point within the image. The dimension of the cell is known as the half-tone screen (e.g. 4 pixels) and defines the real resolution in terms of lines per inch [l.p.i. (d.p.i./half-tone screen)]. Using half-toning also diminishes the digital image depth. The larger the half-tone screen, the greater the number of grey shades, but this must be matched by a substantial gain in d.p.i. resolution to maintain the number of lines per inch (1000 d.p.i is a starting point for a commercial print). An alternative to the half-tone screen is 'dithering', a more random application of dots, which renders gradients more easily, but blurs sharp lines.

In colour printing, laser and ink jet printers use either three or four different inks, each of which is applied with their own independent half-tone screens, or are dithered in such a way as to keep coloured dots separate, thus avoiding moiré patterns. This results in a considerable loss of overall resolution: the d.p.i. value (e.g. 300) is divided by the number of inks used [e.g. four for CMYK (cyan, magenta, yellow, black) prints] and the half-tone screen (e.g. 3 pixels), to give the final resolution (25 l.p.i.). Dye sublimation printers, which melt ink from a three separate ribbon overlays into coated paper, can achieve much higher resolutions and image depth. Pixels tend to blur, so that the resulting image, while appearing 'photographic', is still not as crisp as a good colour print. For publication purposes, it is suggested that at least one set of images be of dye sublimation print quality [4].

Finally, the range of printable colours, known as the 'gamut', depends on the mechanism by which ink is applied to paper. Printed colour cannot match the gamut of colours generated by the cathode ray tube in a colour monitor. Laser and ink jet printers thus have a smaller gamut than that of a computer monitor, while the gamut of dye sublimation printers is even more restricted. The practical consequences are that printed images may appear washed out. Furthermore, a colour that can be distinguished by its increasing saturation on a computer monitor may not be discriminated in print. Some software packages display an `out-of-gamut warning' for particular colour palette selections.

8. Troubleshooting

Problem	Diagnosis
Image appears grainy or noisy	Random noise fluctuations appear when the camera is straining to acquire enough light. The first consideration must be to increase the illumination intensity to its maximum. If intensity is fixed (fluorescent light source) or already maximized, there are a number of options (1) average successive scans of the same image to even out random fluctuations (Section 3.2.5); (2) bin neighbouring pixels to reduce noise, but also reduces resolution (Section 3.2.3); (3) apply *post hoc* 'median' filtering to remove fluctuations while retaining correct, gross image features (Section 5.1); (4) use a cooled detector array, if available (Section 3.1.2).
Unevenness in illumination across the image	Depending on software options, it may be possible to divide the image by a background reference image (Section 3.2.5). Image division results in a reduction in contrast and the field of illumination should be balanced before using image arithmetic. A shading correction digital filter can also be applied to the image *post hoc.*
Colour varies dramatically between successive photomicrographs	Constant voltage is not being applied to the illuminator. On a photomicroscope, there may be a '3200 K' or 'Photo' setting on the camera controller which provides a standardized colour temperature and, therefore, faithful colour reproduction with tungsten films. If no such setting is available, use the light source at its brightest. Some systems will allow a white balance to be calculated for any given colour temperature using a background as reference (Section 3.2.6).
Camera gives a warning that exposure times are too short	Reduce illumination levels using neutral density filters (do not adjust the voltage supply to the light source as this will alter the colour temperature of the illumination).
Image appears washed-out	The full dynamic range of the collector is probably not being used. Optimize gain and black levels by checking the intensity distribution using a colour LUT or a histogram of pixel-value distribution (Section 3.2.1 and Fig. 1).

Image has too much contrast and is saturated at its extremes	Gain and black levels are inappropriately set, resulting in 'clipping' of the brightest and darkest parts of the image. Optimize gain and black levels by checking the intensity distribution using a colour LUT or a histogram of pixel value distribution (Section 3.2.1 and Fig. 1).
Faintest parts of the image are lost as the brightest parts of the image approach saturation	If possible, adjust the linearity (gamma) of the detector array in a positive direction to boost the intensity of fainter pixels (Section 3.2.2 and Fig. 2).
Low intensity background noise detracts from the representation of brighter features within the image	If possible, adjust the linearity (gamma) of the detector array in a negative direction to reduce the intensity of fainter pixels (Section 3.2.2 and Fig. 2). If the background consists of 'shot noise' (single rogue pixels of high intensity) then a median filter can be applied *post hoc* (Section 5.1).
Fluorescence image collected on a CCD camera is obscured	If available, use a confocal microscope system to collect a stack of optical planes through the specimen that can then be 'projected' to give a sharp image (Section 4). Alternatively, the blur can be removed by a deconvolution algorithm, or, as a last resort, a LaPlacian filter.
Final saved image has a quilted appearance and slight loss of sharpness when reopened or printed	Check that primary data have been saved in a 'lossless' as opposed to a 'lossy' file format (Section 6).

References

1. Russ, J. (1995) The Image Processing Handbook, 2nd edn, CRC Press, Boca Raton, FL

2. Minsky, M. (1957) US Patent # 3013467

3. Centonze, V. and Pawley, J. (1995) in Handbook of Biological Confocal Microscopy, 2nd edn (Pawley, J., ed.), pp. 549–569, Plenum Press, New York

4. Saper, C. (1999) J. Comp. Neurol. **412**, 381–382

5. Wingate, R.J.T. (1999) in Molecular Embryology (Sharpe, P.T. and Mason, I., eds), pp. 711–733, Humana Press, Totowa, NJ

Three-dimensional microscopy

**Gary L. Greenberg*
and Alan Boyde**

*To whom correspondence
should be addressed.

Chapter 3

Page

1. Introduction

Microscope specimens are three-dimensional (3D) objects, but conventional research microscopes produce flat, two-dimensional images. As a result, high-power light microscopes have been designed for viewing rather thin sections taken from thicker, real-world specimens. Consequently, there is a loss of information regarding the inter-relationships between structures not present within the section being viewed. This can pose a major problem for biologists attempting to understand the location and function of different molecular events within an organism.

3D

There is a clear and present need to examine thicker specimens, such as whole-mount preparations. However, when using conventional microscopes to examine thick preparations, important information can be lost and data can be misinterpreted. This is because conventional microscopes provide resolution only in the *X–Y* plane. We are unfailingly surprised by how much more we can see and discover by studying microscopic specimens in three dimensions. This fact opens up new vistas for strategies for surveying structures within an organism. Readers might ask: what are the specific fields of application for high-power stereo microscopy? The answer is practically all areas of application in which the microscope is used to examine 3D materials, including almost all specimens except exceptionally thin sections.

Having two views of a specimen enables us to exploit an enhanced depth-of-field by separating features in the vertical axis via binocular stereopsis. In a 3D microscope, a larger depth-of-field does not mean a poorer *Z*-axis resolution, as is the case for a conventional microscope. The distribution of features along the *Z*-axis is rendered observable by the angle of parallax between the left and right views. Consider a specimen containing two small particles, one on top of the other. A conventional two-dimensional (2D) microscope would only show a single particle, but a 3D microscope reveals two, one above the other.

Even beyond the depth of sharp focus, out-of-focus portions of the images have informational value in a 3D microscope, because the out-of-focus structures can be seen to be above or below the structures that are in focus.

A principal aim in using 3D imaging is to avoid misinterpretation through misrepresentation, thereby achieving a better understanding of structure. Therefore, it can be very important for biological researchers to examine thicker specimens, in order to gain a direct understanding of the interrelationships between structures within a single sample.

2. Methodology

Many methods have been developed that allow researchers to examine thick, real-world specimens such as thick sections, whole-mount preparations and whole embryos. Hammond [1] has written a good review of the principles of stereoscopy in high-power microscopes using single objective lenses. In this chapter, several methods will be described for viewing, capturing, recording, measuring and publishing 3D images.

2.1. Direct-view 3D microscopy

Using stereopsis, the human eye–brain complex perceives the world in three dimensions even though the retina of each eye receives only 2D images. The brain can use a number of cues, including convergence, motion parallax and the obscuring of background objects by foreground objects, to create 3D space. When viewing a 3D scene with two eyes, each eye receives an image from a slightly different point of view and the brain then fuses these images into a single 3D image. A similar process is used to create 3D imaging in direct-view 3D microscopes.

When using 3D microscopes it is important that the interocular distance (between the two eyepieces) is set properly so that each eye has a good view of each eyepiece. Furthermore, the eyepieces should be are properly focused so that both the left and right eyes are focused at the same level within the specimen.

Direct-view 3D imaging is achieved by differential viewing of the left and right sides of the objective aperture, simultaneously (or in very quick succession). This can be accomplished in several ways, which are outlined below.

2.2. Polarizing filters

At the rear aperture of the objective lens, the light rays for the left- and right-eye views are distinctly separated at the left and right halves of the objective aperture. This is the location in the optical system where the left and right images can be 'dissected out'. If one can code the light that passes through the left and right portions of the objective aperture and channel the two separate images to the appropriate oculars, then a 3D view of the specimen will be produced directly through the eyepieces. Unfortunately, the rear aperture of the objective lens is inaccessible for the physical placement of an apparatus that would split the light from its left and right halves.

However, there are several ways to overcome this problem. The condenser lens aperture is conjugate with the plane of the objective lens aperture. Therefore, the condenser lens can be modified to code the left- and right-eye illuminating beams. The best known method has been called 'half aperturing'. The condenser lens is fitted with polarizing filters placed at the plane of the aperture. This is an excellent technique for creating 3D images in a transmitted light microscope using a single objective lens.

Fig. 1 shows half of the condenser aperture covered with polarizing material placed in one orientation (northwest by southeast) and the other half of the condenser aperture covered with polarizing material placed at 90° (i.e. northeast by southwest). Another set of polarizing filters is placed in the microscope eyepieces and rotated such that the light from one half of the condenser is blocked from entering one eyepiece, and light from the other half of the condenser aperture is blocked from entering the other eyepiece. This will produce a real stereo image directly via the microscope eyepieces. If the result is reversed stereo (i.e. the foreground appears to be in the background), then swap the orientations of the filters in the eyepieces.

This method usually gives very good stereo imaging with true colour rendition. An added advantage is an apparent enhancement of resolution and depth-of-field in the image. The method works well with most conventional transmitted-light microscopes. A possible problem can occur if the prisms in the head of the microscope depolarize the light.

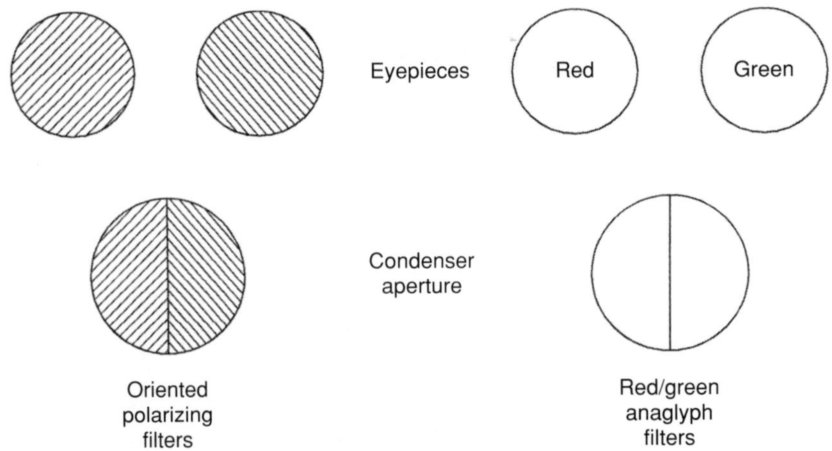

Fig. 1. Using filters in the condenser aperture for creating direct-view 3D imaging in a conventional microscope

This will result in the de-coding of the left and right light paths and the loss of 3D information.

2.3. Anaglyph filters

Another method uses complementary coloured filters to code the left- and right-eye images. Anaglyphs are stereo images that are recorded as a single image rather than as a stereo-pair. The anaglyph technique uses a red filter in the left half of the condenser aperture and a green (or blue or cyan) filter in the right half of the aperture. A red filter is placed in the left eyepiece and a green filter is placed in the right eyepiece in order to see the 3D image directly at the microscope (see Fig. 1). Red and green gelatine filters can be purchased from any photographic store and can be easily cut into the appropriate size and shape to fit into the back of a conventional condenser lens. The filters should be cut into opposing 'D' shapes and placed as close as possible to the rear aperture of the condenser lens. It is advantageous if they overlap slightly in the midline, as this creates a central black strip, from which no 3D information can be obtained because the corresponding 'rays' contribute equally to both eyes. When placing the condenser back in the microscope, be sure to rotate the condenser so that the red filter is on the left and the green filter is on the right side.

3D

Anaglyph filtering is unsuitable for coloured objects, and limits resolution to that imposed by the longer wavelength red light that is used. However, it can be a very effective method for producing 3D images on a video screen, provided that the specimen is not heavily coloured. When using the anaglyph technique for video, an RGB monitor (a colour monitor with separate access to red, green and blue channels) should be used and the blue lead should be unplugged. The 3D image can be seen on the video monitor by using red/green stereo glasses (glasses can be purchased at Reel 3D Enterprises, Inc. or 3-D Images Ltd).

2.4. Shutter mechanisms

Another alternative is to use rapidly alternating shutters. In 1985 Victor Kley [2] invented a computer controlled LCD shutter installed in the aperture of a condenser lens. The shutter could be programmed to quickly block light passing the left or right halves of the condenser aperture. LCD shutters placed in the eyepieces were synchronized with the rapidly alternating half-aperture illumination to produce direct-view 3D images. This system is manufactured by Zeiss as an add-on accessory for their transmitted- and reflected-light microscopes. The system can be synchronized at video rate for recording and viewing on 3D video systems.

2.5. Multiple oblique illumination

Greenberg and Boyde [3,4] developed an illumination system that uses independent, oblique illuminating beams to create high-definition, direct-view 3D images. As the illuminating beams pass through the specimen at different oblique angles, they create apparent tilted views of the specimen, even though the specimen has not moved. Thus, left- and right-tilted oblique illumination creates left- and right-eye views of the specimen. The right oblique illuminating beam will shift the background structures to the right and the foreground to the left, relative to the central structures in the plane of focus. The opposite applies to the left oblique illuminating beam. These shifts are the cues needed by the eye–brain complex to convert the 2D images of the specimen into a single 3D image.

To ensure that each eye only sees the intended oblique illuminating beam, the left and right channels are coded using polarizing filters. Placing complementary filters in the left and right eyepieces allows a single objective lens to create left- and right-eye images of the specimen simultaneously. The method is applicable using a wide range of conventional objective lenses, including phase-contrast optics.

Coincidentally, obtaining stereo images via oblique illumination gives rise to important imaging advantages. The earliest realization of these advantages was noted by Abbe [5], who published the details of a means for producing a single, controllable oblique illuminating beam. Abbe showed that oblique illumination shifts the diffraction pattern relative to the objective aperture, thereby capturing higher order diffraction wavelets within the objective aperture. The practical resolution of the microscope can be improved by careful employment of multiple oblique illuminating beams [3,4].

The most practical benefit of oblique illumination, compared with conventional illumination, is the improvement in contrast and depth-of-field. Depth-of-field and contrast are dependent on the cone angle of the illuminating beam: the smaller the overall cone angle, the greater the contrast and depth-of-field. Oblique illumination can reduce the overall cone of illumination without reducing the numerical aperture (NA) of the optical system. By tilting and positioning the illuminating cone to one side of the optical system, resolution can be enhanced, while adding contrast and depth-of-field to the image. Carefully controlled oblique illumination allows the microscopist to optimize resolution, contrast and depth-of-field. Specimens can be viewed under a wide variety of illuminating conditions, revealing subtle details by creating highlights and shadows. This underscores an important benefit of this technology, namely, the ability to examine living, unstained specimens and other samples that have inherently low contrast.

A multiple oblique illumination system that replaces the standard light source on conventional light microscopes was developed by Edge Scientific Instrument Company. The system provides 3D imaging directly through the eyepieces of standard, high-power microscopes and improves image quality regardless of whether the microscope is used in 2D or 3D modes. The illumination system can be retrofitted to either transmitted-light or reflected-light microscopes, of either the upright or inverted type. This accessory makes high-definition, real-time 3D microscopy affordable and simple to use.

2.6. 3D head method

The Edge Scientific Instrument Company also developed a microscope head that converts the 2D image from a conventional microscope into a real-time 3D image. As with the previous technique, it also employs conventional objective lenses. Direct-view 3D imaging is achieved by simultaneous inspection of the objective lens from left and right viewpoints [3,4]. Polarizing filters or coloured filters are not required, and the head functions well with fluorescence microscopes as well as transmitted- and reflected-light microscopes. Unlike the multiple oblique illumination method, this technique is independent of the type of illumination, and is compatible with a range of optical systems, including transmitted light, phase contrast, DIC (differential interface contrast), darkfield, brightfield, oblique light, incident darkfield, epi-reflection, epi-polarization, transmitted linear and circular polarization, and fluorescence microscopy.

The 3D head method does not increase resolution in X–Y, as does the multiple oblique illumination system. However, the addition of perceivable resolution in the Z-axis produces an image that appears very sharp.

2.7. Capturing and recording stereo images

Capturing stereo images utilizes the same camera equipment used for capturing conventional 2D images, i.e. 35 mm film cameras, digital cameras and video cameras. There are three main ways of viewing stereo images: stereo-pair images, anaglyph images and lenticular images. Each way requires a different method for capturing the images.

2.7.1. Stereo-pair images

Stereo-pair images consist of a left-eye view and a right-eye view of the specimen. Most types of 3D microscopy involve some form of half aperturing, i.e. using half of the aperture for creating a left-eye image and the other half for the right-eye image.

It is important that the left- and right-eye images are taken with the same exposure, in case one of the images appears darker than the other, thereby making it more difficult for the eye–brain complex to fuse the images into 3D.

It is critical that the camera(s) are aligned with the *X–Y* axes of the microscope. The photographic format must be either portrait (vertical image) or landscape (horizontal image), and the recorded images must be viewed in the same rotation in which they were taken. Stereo-pair images are best taken in the portrait format because this reduces the amount of convergence necessary to fuse the images.

The left-eye and right-eye images can be captured either simultaneously with two cameras, or sequentially with one camera. If the specimen or parts of the specimen are in motion, then simultaneous capturing of left and right images will be required. If the specimen is static, then they can be captured sequentially. Fluorescence images, however, should be recorded simultaneously. If they are recorded one after the other, then the left-eye image may look different from the right-eye image owing to fluorescence bleaching.

A dedicated 3D microscope, such as the Edge True-View 3D Head, may have two camera ports, which capture the left- and right-eye images simultancously. If a conventional microscope is being employed, then a single camera can be used and the left and right images can be captured sequentially.

If polarizing filters are employed in the condenser aperture of a conventional microscope, the stereo-pair photos can be captured through two camera ports simultaneously if a microscope that permits the attachment of two cameras is used. Alternatively, a photo accessory that converts a single camera port into a dual camera port can be used. Polarizing filters must be placed in the photographic path of each camera and must be rotated so that one camera sees the left-eye image, and the other camera sees the right-eye image. In order to rotate the filters properly, it will be necessary to block the left half of the condenser lens with an opaque material and rotate the polarizing filter in one of the camera paths so that the image appears as dark as possible. The other half of the condenser lens is then blocked and the filter is rotated in front of the other camera path until it is as dark as possible. The two cameras will now be set to record stereo-pair images simultaneously.

Physical half-aperture stops can be used to record the two members of the stereo-pair sequentially and without the side effects of the anaglyph or polarizing filters, but this method usually precludes 3D viewing through the eyepieces.

A turret-type condenser lens, such as a phase-contrast condenser, can be used to create stereo-pair images, as follows.

1. Place the condenser to the brightfield setting and remove one eyepiece from the trinocular head.
2. While looking down the tube of the microscope, slightly rotate the turret so that the left half of the condenser aperture is obscured (see Fig. 2A).
3. Mark the position of the turret with a pen or pencil so that the precise position can be easily repeated.
4. Rotate the turret slightly, so that the right half of the condenser aperture is obscured (see Fig. 2B).
5. Mark the position so that it can be repeated without having to look down the tube of the microscope. These two settings can now be used to record stereo-pair images in a sequential manner.

6. Simply move the rotating turret to the left position and capture the left-eye view, then rotate the turret to the right position and capture the right-eye view. Although this method does not allow for 3D viewing directly through the microscope, it does provide a simple method for capturing true-colour stereo images on a conventional transmitted-light microscope.

2.7.2. Anaglyph images

To capture stereo images using the anaglyph method, set up the condenser lens with red/green filters as described in Section 2.3. Then, simply record the images with a photographic camera or digital camera. The resulting photos will look like fuzzy red/green images. However, if the photos are viewed with red/green stereo glasses, then a sharp 3D image will appear.

3D

Depending on the type of film being used and the optical density of the red and green filters in the condenser lens, the resulting anaglyph image may have too much red or too much green. If this is the case, then use an appropriate neutral-density filter to produce an even balance of red and green in the final image.

Alternatively, any stereo-pair of images, monochrome or coloured, can be viewed by projecting them on to the same plane through red and green

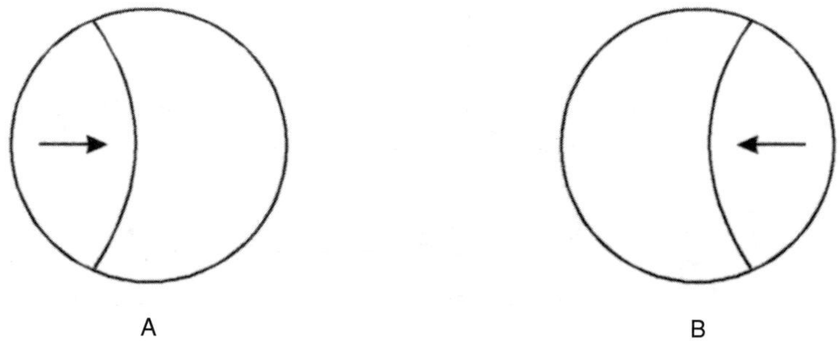

Fig. 2. Using a turret-type condenser lens to capture stereo-pair images

filters, with the viewer using red/green spectacles. It is important to note the international convention that *red* is *port* (left) and *green* is *starboard* (right).

2.7.3. Lenticular images

Lenticular stereo images consist of multiple exposures of a specimen from multiple points of view; for example, left, middle and right. The separate images, representing different angled views of the specimen, are printed onto a single image in such a way that when a plastic lenticular screen (composed of very fine vertical prisms) is properly positioned over the composite image, the component vertical images strips are seen only by the corresponding eyes and a 3D image is created. 3D postcards use this technology. The great advantage of this technique is that no glasses or other viewing aids are required for seeing the stereo images.

Lenticular images can be printed by companies such as 3Dx in the U.S.A. and David Burder in the U.K. 3Dx can print 3D images using either three or four points of view if they are photographed on half-frame 35 mm film. 3-D Images Ltd, and other lenticular specialists, can print 3D images using from two, to say, 36 images, taken from slightly different points of view. The more images that are used, the better the 3D image will appear. The images are known as Integrams and PseudoHolograms when they have many components.

A turret-type condenser lens, such as a phase-contrast condenser, can also be used to create images from three points of view for lenticular printing, as follows.

1. Rotate the turret to the brightfield setting and remove an eyepiece from the trinocular head.
2. While looking down the tube of the microscope, rotate the turret slightly so that the left half of the condenser aperture is obscured (see Fig. 2A).

3. Mark the position of the turret with a pen or pencil so that the precise position can be easily repeated. Do the same for marking the position that obscures the other half of the condenser aperture (see Fig. 2B).
4. To capture images from three points of view, move the rotating turret to the left position to capture the left-eye view, rotate the turret to the middle of the brightfield position to capture the middle point of view, and finally, rotate the turret to the right position to capture the right-eye view.

3. Using confocal microscopes for 3D imaging

In confocal microscopy, the general and original aim was to overcome the difficulties caused by the too great a depth-of-field of conventional lenses, combined with the problems of general light scatter or of fluorescence occurring simultaneously at many levels within the sample. It was soon realized that the optical sectioning potential of confocal methods could be used to extract many separate images of separate focal planes, thereby creating 3D *X–Y–Z* data sets, from which 3D images can be reconstructed, usually by computer means [6–9]. These very strong advantages of confocal microscopy can be matched in some cases by deconvolution of conventional light-microscopy images, for which the 3D point-spread function of the particular objective lens has been, or can be, determined [7].

More comparable with the present studies are the real-colour reflection or fluorescence stereo images generated with tandem scanning confocal microscopes (TSMs), as described by Boyde [8]. In these microscopes, the multiple focal planes selected to comprise the 3D images are 'stacked' directly as stereo pairs on colour film (or on an integrating charge-coupled device camera), while changing the focus along two oblique, inclined paths. The stereo image can only be seen after processing the film [or immediately after downloading the digital CCD (charge-coupled device) images]. Allowing digital control of the *X* (interocular) and *Z* (focus) directional axes of the objective lens with piezo-electric drivers

3D

under computer control, this method lends itself perfectly to generating the multiple-viewpoint projections needed to form perfect integrams (also called pseudoholograms). Digital image recording, with a through-focus pass for each image of only 1 s, gives rise to excellent stereo images, which can be examined within a few seconds on the video display screen. The photo image is, however, preferred for its much higher information content as compared with digital images. Few current confocal microscope systems work with pixel matrices greater than 1024 × 1024, and these make a poor comparison with photographs.

Confocal scanning laser microscopes (CSLMs) store images of one 'optical section plane' at a time. They are the commonest confocal microscope in use today. Every commercial CSLM system comes with a software package that computes new and alternative projections through the resulting *X–Y–Z* data stack. Software companies are thriving on developing specialist packages for this applications area. Advantages in computer processing can include programs that select for only the brightest image features for display along any simulated line of sight through the data stack. Other programs create pseudo-3D representations that show what the object features would look like when cast as a shadow image from one viewpoint. Multiple views can be synthesized and projected in rapid sequence to simulate a tilting or rotating (spinning) object. In such work, however, the Achilles heel of the CSLM will show through: the *Z* resolution is always much worse than that in the *X–Y* plane. For the same reason, the much favoured '*X–Z*' sections from CSLMs always look terrible. In simple terms, the resolution element in a CSLM is a much elongated ellipsoid, i.e. a circle appears like an ellipse when viewed from the side. The lower the NA of the objective and the lower its magnification, the more severe the problem. Eventually, well focused molecular biologists will see the extreme advantages of the instant gratification and WYSIWYG (what you see is what you get) data from the direct view methods (direct view includes direct view confocals, such as the disc scanning 'TSMs'). Real problem solving means getting answers quickly, to know whether there is anything to be seen so that a conclusion can be reached.

4. Measuring 3D images

To measure in the third dimension, it is first necessary to know the parallax angle generated by the objective lens being employed. 3D (X–Y–Z) co-ordinates are obtained by measuring offsets of corresponding features (parallaxes) in the plane of the two images by stereophotogrammetry. Hammond [1] points out that the reason why 3D imaging has not been widely accepted is that it is perceived to be difficult to generate numerical 3D data from such images. As with scanning electron microscopy 3D imaging, where we need to know the tilt angle difference between the two projections making the stereo-pair, we need to know the parallax angle as an indicator of the degree of exaggeration of the stereo depth effect and to make any computations.

In the case of either real-time direct-view or slow confocals, stereo-pair projections are the sums of flat images, i.e. we deal with perfect parallel projection [10]. For the 'see it straight down the microscope' approaches, the projection is a complex perspective one.

We have developed a simple method to determine the parallax angle when using any objective lens under any operational constraint. A calibration slide with 10 μm markings is employed as a test target. The micrometer markings should be either clear, or white on a black background. The microscope condenser lens is used in anaglyph mode, with a red filter covering one half of the aperture and a green filter covering the other half. No colour filters are placed on the eyepieces. When set up in this manner, the out-of-focus structures will appear as red/green double images. The condenser lens must be centred and focused properly, and the condenser iris must be placed in the position in which it will be used when collecting the data images. One should then focus on the markings inscribed on the stage micrometer. The image should appear white on a black background. The operator has to determine how much de-focusing of the stage is necessary to achieve an image that looks like evenly spaced out-of-focus red and green lines. This corresponds to an apparent 5 μm offset between the left and right images of each line in the test specimen (see Fig. 3). The change in focal position is read from the microscope focus dial indicator and the parallax angle is computed by triangulation.

Fig. 4 shows the relationship between parallax angle and the NA of the optical system. The NA of illumination is kept the same as the NA of the objective lens being tested. High-NA objectives produce exaggerated stereo images, or hyperstereo. When the NA increases, the depth-of-field decreases. If the magnification and NA are very high, then the depth-of-field will be very shallow. The exaggerated parallax favours the ability of the eye-brain complex to perceive and analyse small 3D volumes. Therefore, although exaggerated stereo may initially seem undesirable, it is important in order to perceive stereo in such thin optical slices of the specimen.

By closing the iris diaphragm of the condenser lens can decrease the parallax angle. This will reduce the exaggerated stereo effect seen in the image. However, closing the condenser iris by too much will reduce the resolution and sharpness.

5. Presenting and publishing 3D images

There are several ways of presenting and publishing stereo images. There are advantages and disadvantages of each method. Here we will focus on how to view and publish stereo-pair images.

5.1 Presenting 3D images

The easiest and most effective way of viewing stereo images is with a twin 35 mm slide viewer. Stereo slide viewers are the best method for personal viewing of 3D images and it is the best way of sharing these images with a small number of colleagues. The disadvantage is that a group of people cannot view the images simultaneously.

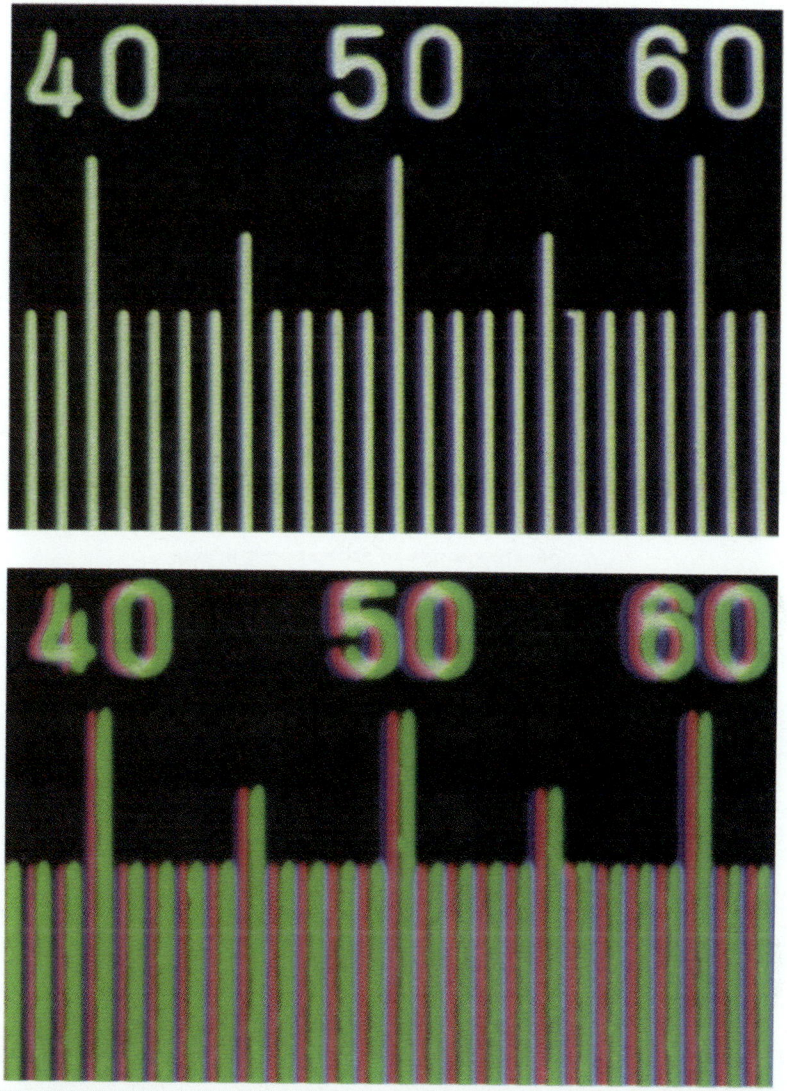

Fig. 3. Using a stage micrometer to determine the parallax angle of a 3D optical system

Upper panel: photomicrograph of micrometer with 10 µm spacing, in focus, using red and green light. Lower panel: the same field of view, defocused to precisely interdigitate the red and green images equally. The change in focus (delta height) represents a 5 µm shift in the X-axis. The slope of the hypotenuse is half the parallax angle.

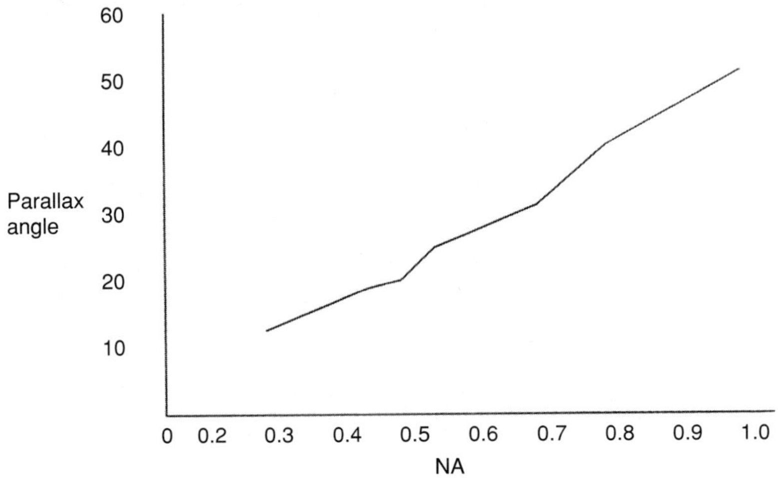

Fig. 4. Direct relationship between the NA of the optical system and the parallax angle of sthe stereo image

Note that higher NA objectives produce exaggerated stereo parallax. Nikon objective lenses tested were: 10/0.25 LWD 160/0; 20/0.4 ELWD 210/0; 10/0.45 PlanApo 160/0.17; 40/0.5 ELWD 210/0; 40/0.65 Plan 160/0; 20/0.75 PlanApo 160/0.17; 40/0.95 PlanApo 160 with coverslip correction.

The classical preferred method for presenting 3D images at seminars and conferences is 35 mm stereo slide projection. The following equipment will be needed: two slide projectors with matching lenses; polarizing filters that are the right size to cover the projector lenses; a silver or lenticular projection screen; and 45°/135° polarizing glasses for the audience. Optional accessories would include a twin projector stand to facilitate projector alignment and a twin remote control in order to advance the slides simultaneously. If a proper silver screen is not available or not large enough, a 3D projection screen can be made by spraying a large flat surface, e.g. the smooth side of 'hardboard' with aluminium paint.

If a twin projector stand is not available, then the two projectors should be placed side-by-side.

1. Load one projector with the left-eye images and the other projector with the right-eye images.
2. Aim the two projectors so that the left- and right-eye images overlap one another as closely as possible.
3. Turn off the lamp on the right-eye projector in order to see the left-eye image only. While wearing a pair of 45°/135° polarizing glasses, rotate the polarizing filter covering the projector lens so that the right eye goes dark. Keep your head level while determining the correct rotation position.
4. Turn off the lamp on the left-eye projector and turn on the lamp on the right-eye projector.
5. Rotate the polarizing filter covering the lens so that the left eye goes dark.
6. Turn on both projectors and a 3D image will appear.

When projecting the slides, it is important to note that alignment to overlap in the vertical direction is the most critical. An assistant should be ready to adjust the vertical position of one of the projectors to correct for any non-matching alignment during the slide presentation. It is easier to perceive any misalignment if the person attending the projectors is not wearing the 3D polarizing glasses. The audience should be told that if they are sitting too far from the centre of the screen the 3D effect will diminish.

Another way of presenting 3D images to a group of people is on video or computer monitors. The basic method is to interleave the left- and right-eye images so that they alternate rapidly enough to fuse into a single 3D image. A multiplexing unit is used to interleave the images, and the use of a 120 Hz monitor is preferable. The viewers must wear 'CrystalEyes', or similar electronic LCD spectacles (available from the StereoGraphics Corporation), or polarizing glasses while employing a synchronized LCD panel covering the monitor.

5.2 Using a flat-bed reflective scanner to generate stereoscopic pair images and processing to anaglyph.

Flat-bed scanners are also microscopes, probably the cheapest known form of scanning optical microscope. At a resolution of 1200 d.p.i., the pixel size is 21 μm, and better can be obtained. The pixel size for resolution by the naked eye is commonly taken to be 75 μm. Remarkable depths of field are obtained with such systems. Placing the same object at different distances from the centre line of the scanner bed gives rise to images in which regions further from the glass plate are projected sideways to different extents. This oblique projection can be harnessed to generate 3D images. Here we outline a procedure based on the use of a UMAX UTC 5400 colour scanner, with image capture and processing in Paint Shop Pro 5 (or later).

1. Place the object on a thin glass plate with an edge contacting the top edge of the available frame 3 cm to the left of the midline of the scanner.
2. Perform a preview scan and select a field encompassing the area of interest.
3. Set the resolution as required and scan.
4. Move the object with the plate (to ensure parallel movement and thus avoiding any rotation of the object) to 3 cm to the right of the midline. Preview, select an area and scan again.
5. Adjust the sizes of the images so that they are the same. Note the width and height of the left image and set the size of the right to be the same using the CANVAS SIZE option.
6. Use the cursor to find the co-ordinates of a feature near the centre of the field of view in the left and right images.
7. Using the CANVAS SIZE option, shift the right image by adjusting the TOP and LEFT options, keeping the canvas size the same.

8. Check that the chosen feature now has the identical pixel address in both images.

   Go to COLOURS, CHANNEL COMBINING, and COMBINE FROM RGB. Choose LEFT (the name of the left image) as the red channel source, RIGHT (the name of the right image) as the green channel source and RIGHT again (the name of the right image) as the blue channel source and combine.

   The resulting anaglyph stereo pair can be viewed on a monitor and will project beautifully using standard data projectors, such as PowerPoint. Red–green or red–cyan viewing glasses can be used.

5.3 Publishing 3D images

Publishing 3D images in journals is achieved by printing the left- and right-eye images side-by-side. A trained reader can free-view the two images and fuse them into a single 3D image. Many scientists in disciplines such as chemistry and crystallography routinely use this method. Nearly every issue of *Science* and *Nature* includes stereo-pair images showing the 3D structure of complex molecules. The reason that the stereo-pair images are published is because they allow the viewer to perceive complex 3D structures that cannot be fully conveyed or appreciated in 2D.

There are two main techniques for free-viewing stereo-pair images: parallel (wall-eyed) viewing, and cross-eyed viewing (which gives a pseudoscopic inversion of depth/height distributions). Parallel viewing is the standard form. The images are printed side-by-side with the left-eye image placed on the left and the right-eye image on the right. The width of each print should be no greater than 63 mm because that is the average distance between human adult eyes (interocular distance). However, the dimension of the prints in the vertical direction is unrestricted. It is critical that the prints be placed so that the corresponding features in the

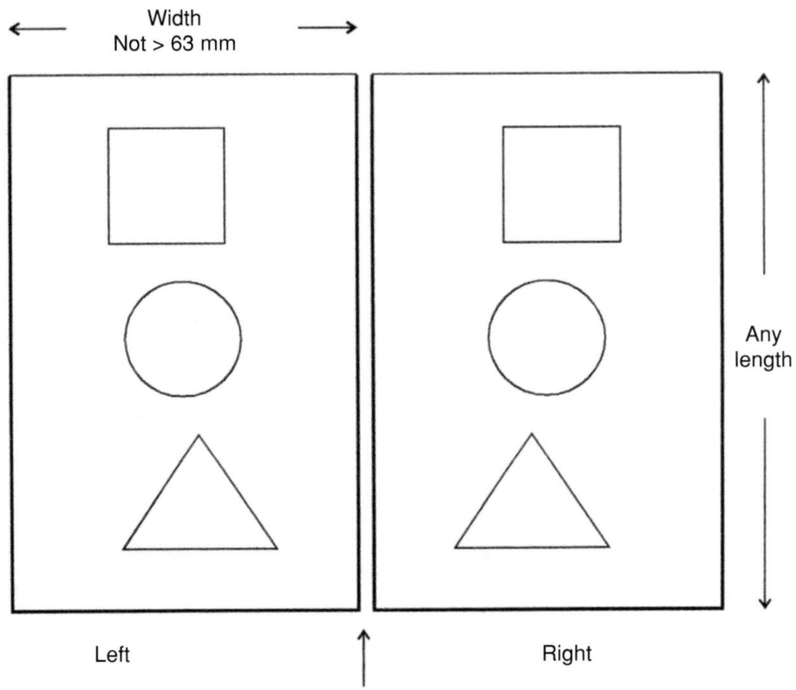

Fig. 5. Presentation of stereo-pair images for publication in journals

image are in the same horizontal plane (see Fig. 5). The key to fusing the two images is for the observer to disassociate the tendency to cross their eyes when focusing close-up. The following technique is suggested for those who are just learning to view 3D images using the parallel gaze method (see Fig. 6).

1. Place the images very close to your face so that the left eye is directly in front of the left image and the right eye is in front of the right image.

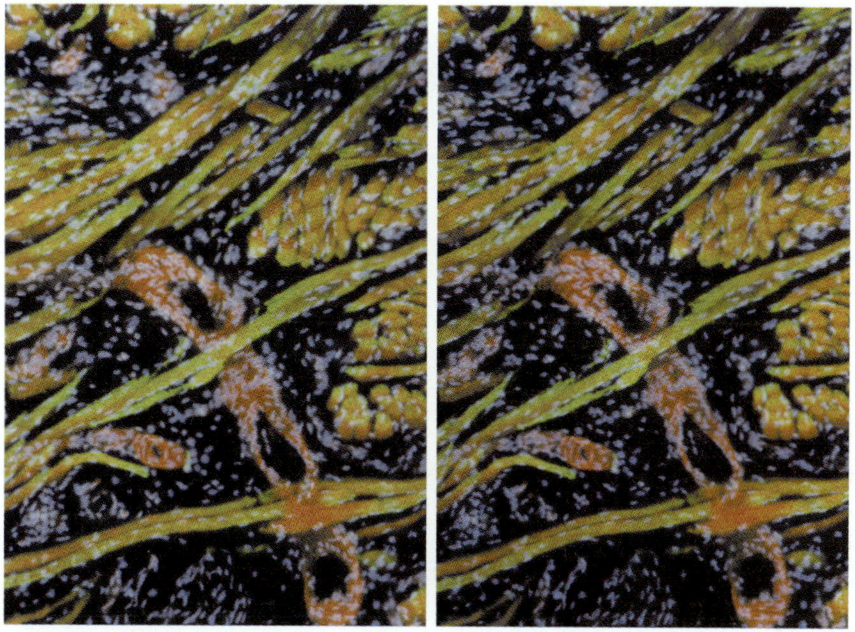

Fig. 6. A stereo-pair photomicrograph showing triple-labelled fluorescence in mouse tongue

Red, actin in blood vessel; yellow, actin and myosin in muscle; blue, DNA. Magnification 125×.

2. While staring at the images as though you were gazing at a blank wall in the distance, slowly move the images away from your face to a comfortable viewing position.

3. When three images appear, concentrate on the centre one until it becomes 3D.

   It is worth spending some time trying to learn this technique. It will open up a world of appreciation for, and understanding of, 3D space.

   If free-viewing of the images proves too difficult, a viewing aid, such as a lunette, can be used. This consists of a pair of magnifying glasses with prisms that redirect the gaze of the right and left eyes for parallel viewing, counteracting the tendency to cross the eyes when focusing close-up. For details see the section on suppliers and resources to purchase 3D viewing aids and other 3D accessories.

References

1. Hammond, C. (1996) Quekett J. Microsc. **37**, 566–574

2. Kley, V.B. (1985) U.S. Patent No. 4,561,731

3. Greenberg, G. and Boyde, A. (1997) Microsc. Anal. **22**, 7–9

4. Greenberg, G. and Boyde, A. (1997) Proc. R. Microsc. Soc. **32**, 87–101

5. Abbe, E. (1873) Arch. Mikroskop. Anat. **9**, 469–480

6. Pawley, J.B. (ed.) (1995) Handbook of Biological Confocal Microscopy, Plenum Press, New York

7. Fay, F.S., Carrington, W. and Fogarty, K.E. (1989) J. Microsc. **153**, 133–149

8. Boyde, A. (1985) Science **230**, 1270–1272

9. Boyde, A. and Jones, S.J. (1995) in Handbook of Biological Confocal Microscopy, 2nd edn (Pawley, J.B., ed.), pp. 255–266, Plenum Press, New York

10. Howell, P.G.T. and Boyde, A. (1984) in The Analysis of Organic and Biological Surfaces, (Echlin, P., ed.), pp. 325–349, John Wiley and Sons, New York

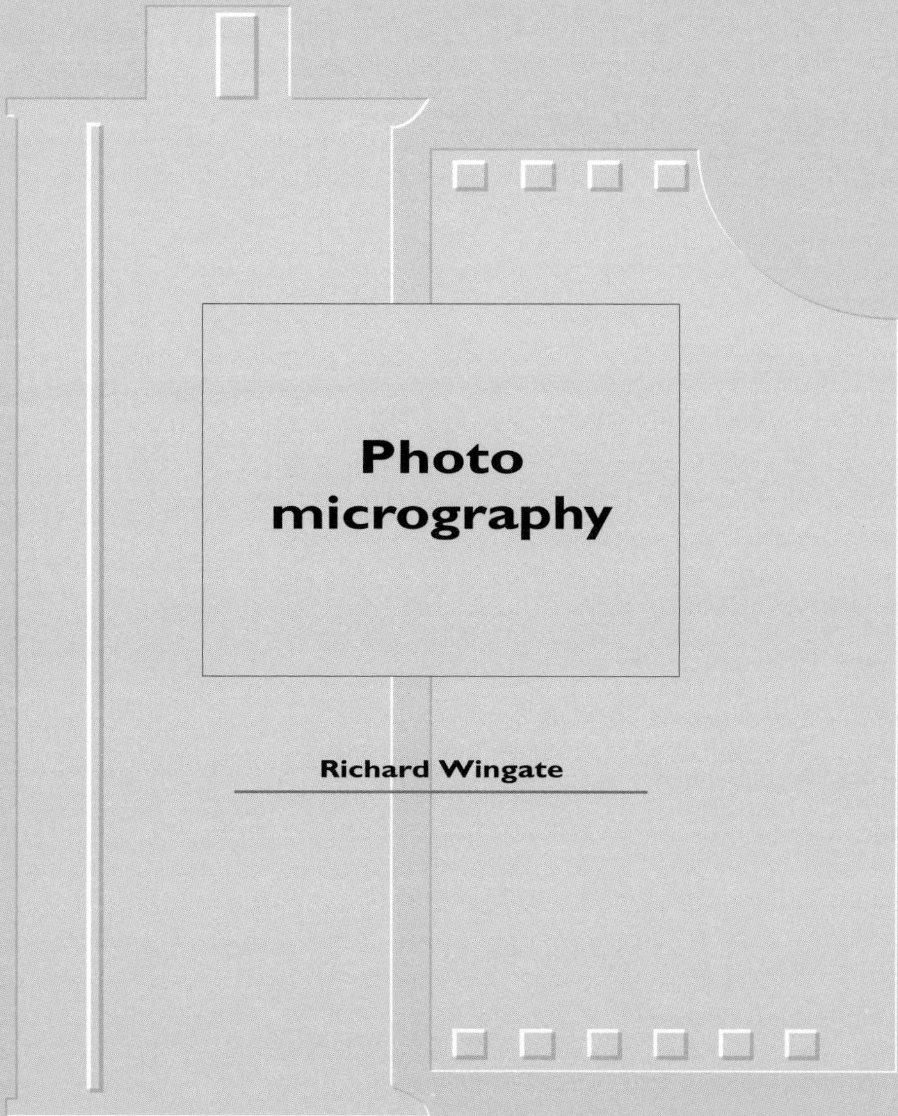

Photo micrography

Richard Wingate

Chapter 4

1. Introduction

Photomicrographs fulfil a number of roles in biological research. Their primary purpose is often simply to illustrate material in presentations or publications supporting results otherwise described in text. Alternatively, with images of live material or of non-permanent preparations, photomicrographs may become the primary data. As a potential substrate for analysis (e.g. distance, optical density, or complexity measurements), the latter form of photomicrograph should represent a uniform and unbiased representation of the original specimen. Understanding the factors that influence image capture may become critical, not only for a representative illustration of material, but also for the correct analysis and interpretation of data.

The primary limiting factor on producing good photomicrographs is the correct alignment of the microscope (Chapter 1 and Wingate [1]). Thereafter, image acquisition is a question of focusing what is viewed through an eye-piece onto the camera and utilizing the properties of the capture medium (film or digital capture array) to represent reliably the important features of the projected image. While digital image capture (Chapter 2) is increasingly popular for its convenience and negligible processing cost, the resolution of film is still superior to that of most electronic images. Transparencies, in particular, offer a convenient, high-quality and relatively stable storage medium for primary data.

2. Photography through a stereo dissection microscope

Each eye-piece of a stereo dissection microscope is served by what is, effectively, a separate microscope tube. For photography, the camera is directed down one of these light paths and views the specimen from a slightly oblique angle. The long working distance of the objectives inevitably leads to a lower image resolution than is obtained with a compound microscope equipped with low magnification objective lenses. However, for larger, opaque specimens, the possibility of both epi-illumination by fibre-optic light sources and of greater depth-of-field afforded by stereo-microscope objectives offers some distinct advantages. Beyond these adaptations, the principles of image capture are the same as for compound microscopy.

3. Photography through a compound microscope

3.1. Specimen preparation

Material should be cleared to a suitable degree and mounted in an appropriate permanent (setting) or temporary medium. Care must be taken to reduce the number of refractive-index interfaces in the light path, each of which will generate light scattering. For example, specimens should be as close to the coverglass as possible so as not to sandwich a layer of mountant along the light path to the specimen (particularly where dry objective lenses are used).

3.2. Coverglass thickness

Objective lenses are corrected for spherical aberration caused by a coverglass thickness of 0.17 mm (occasionally 0.18 mm). For objective lenses with a numerical aperture of less than 0.3 (typically of low magnifying power), deviations from this standard thickness are not important. Where the numerical aperture lies between 0.5 and 0.65, and for all immersion objective lenses, a variation of ±0.05 mm is acceptable. The optical correction of the highest-quality dry lenses, with numerical apertures between 0.70 and 0.95, requires an optimal coverglass thickness of 0.17±0.01 mm.

3.3. Choice of objective

Photomicrography places special demands on the quality of light transmission through the objective lens. Objective lenses vary in the extent to which they are corrected for curvature of field and chromatic aberrations. The prefix 'plan-' indicates an objective that can provide a flat field of view — a basic requirement for photomicrography. The degree of chromatic correction increases with lens quality and price. It not only determines the faithfulness of colour representation but also image quality. Achromat lenses have limited chromatic and spherical correction. If the specimen is illuminated with multiple wavelengths of light (i.e. white light), coloured fringes may appear at edges within the specimen image. These fringes will have a fuzzy appearance in black-and-white photography. Illumination with a single wavelength of light (i.e. green) will considerably sharpen such a monochrome image.

Semi-apochromatic/fluorite lenses have a higher degree of correction, which in turn gives a higher attainable numerical aperture.

Apochromatic lenses may have up to 18 separate elements to achieve a chromatic correction for red, green and blue wavelengths. These are the optimal lenses for colour photomicrography, but will also give the highest absolute resolution due to their superior numerical aperture.

3.4. Choice of immersion medium

High-quality, high-magnification objectives with a numerical aperture of greater than 1.0 require an immersion medium (between coverglass and lens). The type of immersion medium is usually engraved alongside the numerical aperture on the lens casing (oil, glycerol or water). The refractive index of the lens immersion medium should be matched to that of the specimen mountant. Oil immersion (used with permanent, setting mountants) offers the highest refractive index (nearest to that of the glass) and, therefore, the highest achievable numerical aperture and resolution. It should be noted that unless an immersion condenser lens is also used, the actual numerical aperture achieved is only equal to 1.0 (air), regardless of the optimum numerical aperture engraved on the objective casing. The achievable optimum numerical aperture of a high-quality dry objective is 0.95.

After focusing the objective on the specimen, the condenser must be optimized for Köhler illumination (Chapter 1), by first focusing the sub-stage condenser on the edges of the field diaphragm of the illuminator, and then optimizing the aperture diaphragm of the condenser to match that of the objective. Poor alignment will reduce resolution or may even introduce 'false resolution' artefacts when the aperture diaphragm pinhole is incorrectly set below a critical diameter.

4. Parfocality and focus

Regardless of the sophistication of the photomicrographic equipment, the image viewed through the eye-piece must be in register (parfocal) with the image formed within the camera, or the camera must be focused separately. Dedicated photomicroscopes are designed so that the camera will be parfocal when the diopter of the eye-pieces is correctly adjusted for each user. Camera bodies mounted on t-mounts are less likely to be parfocal and must have their own provision for focusing via the camera eye-piece or a focus telescope. Even on a correctly aligned parfocal photomicroscope, at low magnifications (less than $\times 10$ objective), a focus telescope should be employed to overcome the tendency for the eye to accommodate to slight blurring, which in turn leads to failure to determine the precise focal distance.

5. Photography with black-and-white film

Black-and-white film still has a primary role in photomicrography, owing firstly to its flexibility (greater contrast control, exposure latitude and resolution), and secondly to the often prohibitive cost of colour micrograph publication. It allows the use of colour filters to enhance contrast and (as described above) remove residual chromatic aberration in otherwise disadvantaged, cheaper objective lenses. Where there is no colour contrast in a specimen, such as for fluorescently labelled material, black and white film gives better resolution than colour photomicrography. Generally, monochrome images are recorded to negatives, although black-and-white reversal (transparency) film can also be used.

5.1. Film type selection

Emulsions vary in the broad range of wavelengths to which they are sensitive, the overall speed with which they react to light, and the gradient of the linear part of their response curve (known as gamma) defining their contrast sensitivity. In general, a professional grade, panchromatic (broad range spectral sensitivity), fine grain film (ASA 64) with moderate contrast produces best results. Faster film speeds (ASA 400+), required with fluorescent material, tend to give grainier results but, if optical resolution is a limiting factor, this may not result in perceptibly lower quality of image, unless a portion is enlarged. With respect to contrast, gamma values greater than 1 denote a 'hard' response, while those less than 1 give softer contrast. Hard contrast emulsion might be preferred where a specimen is weakly differentiated; however, any colour contrasts can also be utilized to generate intensity contrast within a monochrome image (see below).

5.2. Colour filters

Differences in colour can be converted to differences in grey-scale by the use of appropriate filtering. For example, if maximum contrast with the background is required, a filter that absorbs, or is complementary to, the colour of the specimen should be used.

Complementary colours are:

- orange: blue
- red: blue–green
- purple: green
- violet: greenish–yellow
- indigo: yellow

Filters can also correct the imperfections in the spectral sensitivity of an emulsion. Panchromatic films are not necessarily uniformly sensitive to different wavelengths of light (particularly deep red). With tungsten (orange) illumination, a corrective, yellow–green filter will thus give the most faithful reproduction of grey shades.

5.3. Printing monochrome positives

Prints for publication require a high degree of contrast and full use of the tonal range between black and white. For this reason, grade 4–5 photographic paper is preferred (as opposed to grades 2–3, which is normally used in commercial processing). This allows grey-shade values to be adjusted such that the darkest part of an image is black and the lightest part is white. Monochrome processing is straightforward and provides the opportunity for the image production to be regulated by the experimenter rather than by a commercial printer.

6. Photography with colour film

Monochrome film cannot substitute easily where colour differences are the biologically relevant information in a specimen. In addition to issues of resolution and sensitivity, colour balance and consistency present particular problems for colour photomicrography. These can be overcome by understanding how emulsion responds to different kinds of specimen illumination, allowing exposure and wavelength corrections to produce consistent results. Most variability can be removed by using a constant illumination voltage and only varying the light passing to the specimen by addition of neutral density filters.

6.1. Choice of film

While colour negative offers similar advantages of exposure latitude as black-and-white negative film, colour reversal (transparency) film is generally preferred for quality, stability of pigments, and speed, e.g. Kodak P1600 can be rated at ASA 400–ASA 3200 (and processed accordingly). The chief disadvantage of the colour negative lies in the need, in most cases, for commercial print processing, where the unusual contrasts and colours of photomicrographs may confuse automatic processing devices designed for daylight subjects.

6.2. Colour temperature and light balancing filters

Artificial light sources provide a more orange/red colour of illumination than daylight. This difference can be quantified in terms of colour temperature (degrees K), and colour film emulsion is designed to respond to either daylight and shorter exposures, or artificial/tungsten (T) light sources and long exposures. While daylight films are used for fluorescence photomicrography, tungsten-rated films are used preferentially for brightfield photomicrography. The latter are chromatically balanced for colour temperatures of 3200 K (type B) or 3400 K (type A). Deviation from these defined wavelengths will change the colour of the recorded image. For this reason, photomicroscopes are generally equipped with a 'photo' or '3200/3400 K' setting, which bypasses the illuminator control to provide a constant, appropriate colour temperature for photography. Since the voltage supply to the illuminator lamp is no longer used to adjust light intensity, neutral density filters in the light path are used to regulate intensity, without changing the wavelength properties, of the lamp output. If a daylight film is being used, a compensating blue filter can be added to the brightfield light path to raise the effective colour temperature of the tungsten light source to the appropriate 5500 K (see Delly [2]).

6.3. Long exposures and reciprocity failure

Failure of the reciprocity law (change in film sensitivity) occurs during long exposure and has a particular impact for colour films. The effective slowing of film speed is compensated for by an increase in the exposure time beyond expected values. Films are supplied with reciprocity values that can be programmed into the camera systems of photomicroscopes or denoted as exposure compensation in 'stops'. Because each pigment has slightly different characteristics, colour films may also change their chromatic properties with long exposure times. Appropriate colour filters can, to a certain extent, compensate and details are supplied with films. After 8 s, only a few films, e.g. Ektachrome 400 ASA, daylight, can be compensated [3,4].

6.4. Colour print production

Colour prints made directly from transparencies tend to gain contrast over the original. Alternatively, colour reversal positives can be used to make a *de novo* set of negatives, which can then be used to generate prints using the original transparency to correct contrast.

6.5. Converting to digital format

For publication, colour prints will ultimately be scanned into a computer. However, figures or slides are, more commonly, being scanned and composed by the experimenter and composite figures submitted directly to the printer. An image scanned at $\geqslant$300 d.p.i. will appear to the same subjective quality in print if reproduced to its original (photographic) size. If the image is to be enlarged (as might be expected for most 35 mm transparencies), the scanning resolution must be correspondingly higher or a deterioration in quality will occur. Scanners may introduce two kinds of error: 'striping', generated by inconsistencies in performance between detectors in a row; and non-uniform illumination both, within images (lower intensities at image edges) and between successive scans.

7. Troubleshooting

Problem	Diagnosis
In brightfield microscopy, coloured fringes at the edges of the specimen appear leading to blurring of the photomicrograph	Coloured fringes arise with achromatic (uncorrected) lenses. They can be avoided by using an apochromatic lens or by limiting the wavelength of incident light, for example by inserting a green filter. Sharp images can then be captured on black-and-white film (see Section 3.3)
Photomicrographs appear out of focus even though the image is sharp through the eye-piece	Eye-piece and camera are not 'parfocal' and hence see slightly different focal planes of the specimen. On some photomicroscopes, eye-pieces can be aligned to the camera by focusing on a 'swing-in' eye-piece graticule or cross-hairs. On others, it will be necessary to focus on the specimen through the camera viewfinder or a separate camera-focusing telescope (see Section 3).
Colour varies dramatically between successive photomicrographs	Constant voltage is not being applied to the illuminator. On a photomicroscope, there may be a '3200 K' or 'Photo' setting on the camera controller which provides a standardized colour temperature and hence faithful colour reproduction with tungsten films. If no such setting is available, use the light source at its brightest (Section 5.2). If the problem arises with long exposures, film may be suffering reciprocity failure (Section 5.3).
Photomicrographs all appear orange	Ensure that you are using a film designed for artificial light sources (tungsten) or if using a standard, daylight film apply a corrective blue filter. If using a tungsten-rated film ensure that light source is sufficiently bright (Section 5.2).
Camera gives a warning that exposure times are too short	Reduce illumination levels using neutral density filters (do not adjust the voltage supply to the light source as this will alter the colour temperature of the illumination).

References

1. Wingate, R.J.T. (1999) in Molecular Embryology (Sharpe, P.T. and Mason, I., eds), pp. 711–733, Humana Press, Totowa, NJ

2. Delly, J. (1988) Photography Through the Microscope, 9th edn, Eastman Kodak Company, Rochester, NY

3. Langford, M.J. (1979) Basic Photography, 2nd edn, Focal Press, London

4. Langford, M.J. (1980) Advanced Photography, 4th edn, Focal Press, London

Fluorescent staining of living cells

Iain Johnson

Chapter 5

Page

1. Introduction

One of the major advantages of fluorescence detection in biological imaging and analysis is the capacity for examining living cells, tissues and organisms [1]. This capacity allows dynamic processes such as signalling, excitability, transport, apoptosis and development to be investigated in more biologically realistic contexts than the disassembled model systems employed in traditional biochemical analysis. While operational requirements such as temporal and spatial resolution are instrumentation-dependent, the selectivity, and to some extent the sensitivity, of the detection system are determined by the design and performance of fluorescent probes [2]. A fluorescent probe is a molecule that is designed to interact specifically with a target in a complex biological specimen and to register the interaction by generating a detectable fluorescence signal. Targets of interest include ions, molecules, organelles and cells. Examples of fluorescent staining techniques for living cells applied at both the multicellular (Method 1) and subcellular (Method 2) levels are described in this chapter.

Fluorescence detection in living specimens presents three technical challenges that are not encountered in fixed-cell or model systems. First, and most importantly, the detection process should not significantly perturb the viability or physiological function of the specimen. Fluorescence detection can produce significant cytotoxicity, either via photochemical reactions mediated by the light required for fluorescence excitation [3,4] or from direct effects, such as respiratory inhibition, mutagenicity or the release of toxic breakdown products, that stem from the intrinsic chemical properties of certain probes [5,6]. Secondly, fluorescent probes are xenobiotics and are subject to metabolism and active transport by living cells. Although potentially problematic in terms of maintaining stable probe concentrations during experiments that may run for several hours, these effects can also be exploited to form the basis of cell function assays. For example the cell viability indicator calcein acetoxymethyl ester (AM) (see Method 1) is extruded by P-glycoprotein transporters, resulting in characteristically weak fluorescent labelling of multidrug-resistant cells [7]. Finally, the plasma membranes of mammalian cells, the cell walls of bacteria and the physical thickness of tissues all present significant obstacles to probe delivery into living specimens. Furthermore, viability considerations dictate that invasive delivery techniques should be avoided if possible. Fortunately, all three challenges are mitigated to some extent by the inherent sensitivity of fluorescence detection, as a result of which the concentrations of probes that have to be introduced are typically very small.

2. Methodology

Method 1. Two-colour cell viability staining of mammalian cells

Existing fluorescent probe methods for assessment of cell viability use differing criteria to determine whether a cell is alive or dead (Table 1). To control for method dependence, two assays with different assessment criteria can be run in parallel on the same cell sample. To implement this in practice, two probes that stain live and dead cells in different fluorescent colours and in a mutually exclusive manner are selected. The fluorescence emanating from the sample can then be easily resolved, with respect to emission wavelength, into two components proportional to the number of live cells and the number of dead cells. Such wavelength-resolved multiplex detection schemes are readily implemented on a wide variety of instrument platforms, including fluorescence microscopes, flow cytometers and microplate readers.

In the example described here, calcein AM (Table 1) is used to produce green-fluorescent staining of live cells (Fig. 1). Calcein AM is particularly suitable for this purpose because of its capacity for passive uptake by living cells and its benign effects on normal cellular function. Ethidium homodimer-1 (EthD-1) (Table 1) produces red-fluorescent staining of

Table 1. Fluorescence-based cell viability assays

Probe	Assessment criteria
Nucleic acid stains (propidium iodide, EthD-1, SYTOX Green)	Interaction of extracellular dyes with nuclear DNA is blocked by intact plasma membrane. Dye fluorescence is enhanced upon DNA binding. Fluorescence is therefore inversely proportional to membrane integrity.
Calcein AM	Conversion of non-fluorescent calcein AM to fluorescent calcein by intracellular esterases (Fig. 1). Retention of polyanionic calcein indicates membrane integrity.
Bis-oxonol [$DiBAC_4(3)$]	Intracellular accumulation of fluorescent dye is membrane potential-dependent.
Alamar blue	Reduction of non-fluorescent dye to a fluorescent resorufin by mitochondrial dehydrogenases.
EuDTPA	Fluorescent lanthanide chelate loaded into cells and released into supernatant when membrane lysis occurs.

EuDTPA, Europium: diethylenetriaminepentaacetic acid.

necrotic cells. EthD-1 is selected because its fluorescence emission spectrum is well resolved from that of calcein (Fig. 2) and its high DNA binding affinity results in improved detection sensitivity relative to functionally similar dyes such as propidium iodide. Furthermore, EthD-1 is approx. 75% larger (in terms of molecular mass) than propidium iodide, decreasing the probability of false positive results due to passage of the dye through small membrane pores.

Materials

- Hanks buffered salt solution (HBSS)
  - 5.4 mM KCl (0.40 g/l)
  - 0.3 mM Na_2HPO_4 (0.045 g/l)
  - 0.4 mM KH_2PO_4 (0.06 g/l)
  - 4.2 mM $NaHCO_3$ (0.35 g/l)
  - 137 mM NaCl (8 g/l)
  - 5.6 mM D-glucose (dextrose) (1.0 g/l)

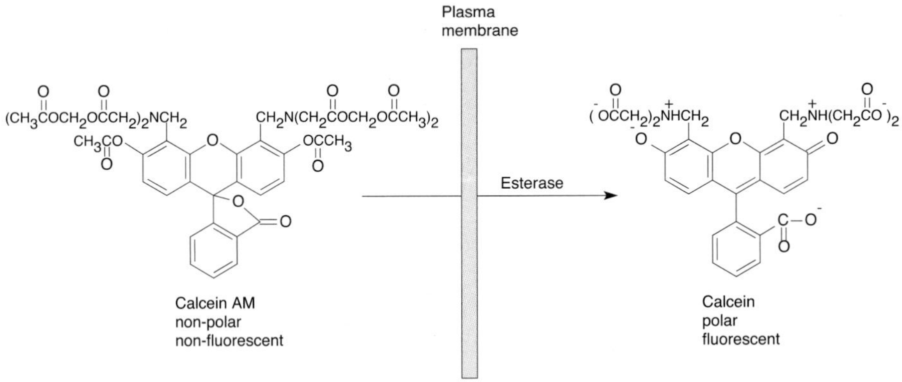

Fig. 1. Principle of calcein AM staining of live cells

Cleavage of the AM ester substituents by intracellular esterases simultaneously modifies both the membrane permeability and the fluorescence of the dye.

Add water to make 1 litre and adjust pH to 7.4 using 1 M HCl or 1 M NaOH if necessary. Autoclave or filter to sterilize.

Note 1 *Any standard saline buffer may be used throughout this protocol. Coloured additives such as Phenol Red should be omitted because of potential fluorescence quenching effects.*

- Calcein AM stock solution
  To obtain a 4mM solution in anhydrous dimethyl sulphoxide (DMSO) dissolve 1mg of calcein AM in 250ml of anhydrous DMSO. Store unused portions of stock solution in microfuge tubes, placed in a container together with a desiccant, at −20 °C.

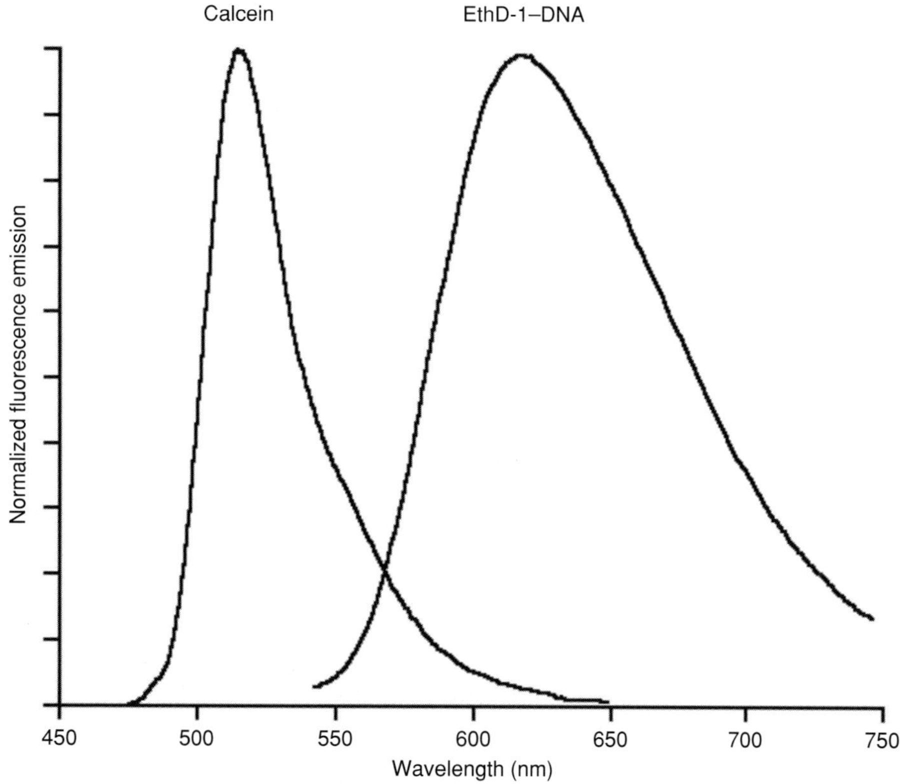

Fig. 2. Normalized fluorescence emission spectra of calcein and DNA-bound EthD-1 in aqueous solutions

- EthD-1
  To obtain a 2 mM EthD-1 solution, dissolve 1 mg of EthD-1 in 480 μl of deionized water mixed with 120 μl of DMSO. Store unused portions of stock solution in microfuge tubes, placed in a darkened container, at −20 °C.

> ***Note 2*** *These solutions are supplied ready-made in the LIVE/DEAD. Viability/Cytotoxicity Kit (Molecular Probes, Inc., Eugene, OR, U.S.A.).*

Equipment

- Fluorescence microscope equipped with a suitable objective (e.g. water immersion 40× magnification, 0.75 numerical aperture) and a long pass filter set designed for fluorescein isothiocyanate (FITC) such as Omega Optical XF115 or Chroma Technology 41012. Alternatively, a filter set with separate emission windows for green and red fluorescence signals (e.g. Omega Optical XF26 or Chroma Technology 71010) can be used.
- Sterile glass coverslips (18 mm × 18 mm square or 25 mm circular diameter)
- Sterile 35 mm Petri dishes
- Glass microscope slides

Procedure

1. Add 20 μl of 2 mM EthD-1 stock solution to 10 ml of sterile HBSS, vortexing to ensure thorough mixing. This gives an approximately 4 μM EthD-1 staining solution.

2. Combine the reagents by transferring 5 μl of 4 mM calcein AM stock solution to the 10 ml EthD-1 solution. Vortex the combined staining solution to ensure thorough mixing.

3. Wash cells prior to the assay to remove or dilute esterases generally present in serum-supplemented growth media (serum esterases could cause some increase in extracellular fluorescence by hydrolysing calcein AM). Gently wash confluent adherent cell monolayers grown on sterile glass coverslips with 200 μl of HBSS (three times).

4. Add 100–150 μl of the combined calcein AM/EthD-1 staining solution (2 μM calcein AM and 4 μM EthD-1) to the surface of a coverslip, so that all cells are covered with solution (see Note 3). Incubate the cells for 20–40 min at room temperature in a covered dish (e.g. 35 mm disposable Petri dish) to prevent contamination or drying of the samples.

5. Following incubation, spot approx. 10 μl of fresh HBSS on to a clean microscope slide. Using fine-tipped forceps, carefully (but quickly) invert and mount the wet coverslip on the microscope slide. Do not apply pressure. To prevent evaporation, seal the coverslip to the glass slide (e.g. with clear fingernail polish).

6. View the labelled cells under the fluorescence microscope. When viewed with an FITC long-pass filter set, green-fluorescent viable cells and red-fluorescent dead cells should be clearly distinguishable within the same field of view.

> ***Note 3*** *The final concentration of DMSO is $\leq$0.1%, a level that is generally harmless to most cells. Note that aqueous solutions of calcein AM are susceptible to hydrolysis. Aqueous staining solutions should therefore be used within 1 day.*

Method 2. Activity-dependent staining of mitochondria

Actively respiring mitochondria generate a large negative transmembrane potential that results in the selective uptake of various cationic dyes. Among these activity-dependent mitochondrial stains, JC-1 (5,5′,6,6′-tetrachloro-1,1′,3,3′-tetraethylbenzimidazolylcarbocyanine) is unique in that its potential-dependent accumulation is indicated by a fluorescence emission wavelength shift (Fig. 3) from green (525 nm) to red (590 nm). The potential-sensitive fluorescence shift of JC-1 is due to concentration-dependent aggregation of the dye [8,9]. The response of JC-1 fluorescence to mitochondrial polarization superficially resembles that of calcein AM/EthD-1 to cytotoxicity (Method 1), although the underlying physical basis is quite different. This similarity allows the two methods described here to be performed using the same microscope configuration. The ratio of green to red JC-1 fluorescence intensities is dependent only on membrane potential and not on other factors, such as mitochondrial size, shape and density, that may influence single-component fluorescence signals. Using green to red ratio detection, membrane potential can be compared within populations of mitochondria and the percentage response to applied stimuli can be assessed. Subtle heterogeneity in cellular responses can be discerned in this way [8]. Currently, the most widely implemented application of JC-1 is for detection of mitochondrial depolarization occurring in the early stages of apoptosis.

Materials

- HBSS
- HBSS supplemented with 10% foetal calf serum
  Mix nine parts of HBSS with one part foetal calf serum.

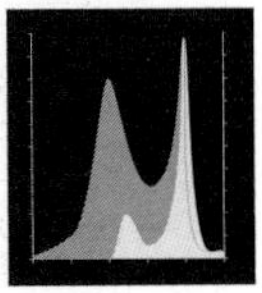

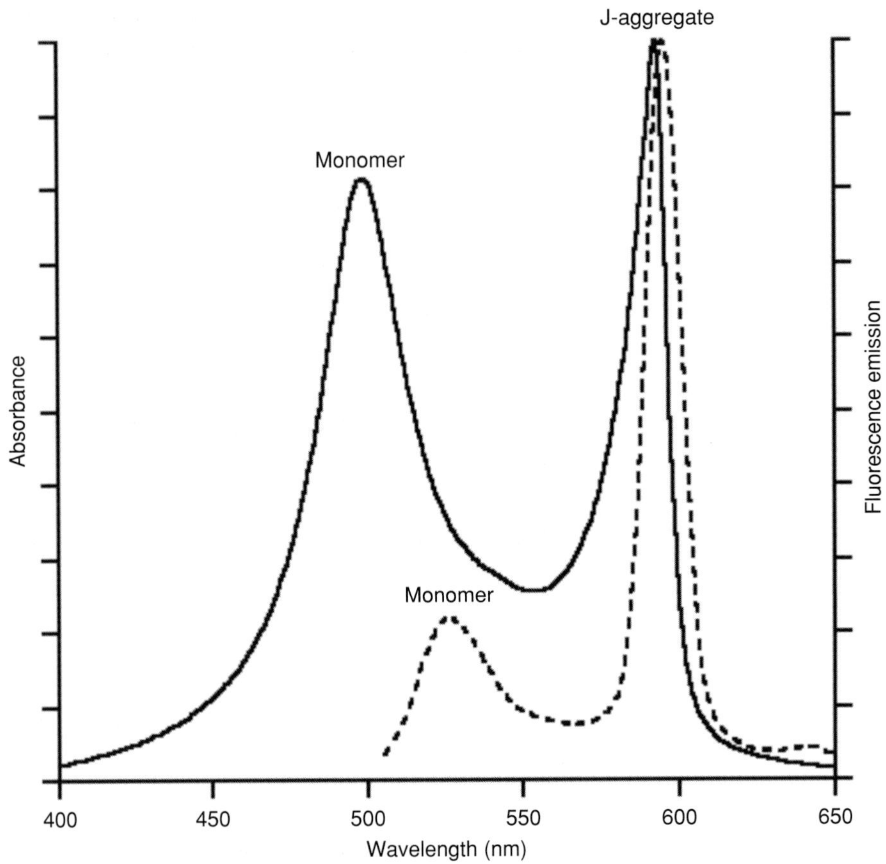

Fig. 3. Absorption (—) and fluorescence emission (- - -) spectra of JC-1

The spectra represent a JC-1 concentration of 10 μM in 50 mM Tris/HCl (pH 8.2) containing 1% (v/v) DMSO. Under these conditions both spectra exhibit two peaks, assigned to the monomer and aggregated ('J-aggregate') forms of the dye.

Note 4 Any serum-supplemented saline buffer may be used throughout this protocol. Coloured additives such as Phenol Red should be omitted because of potential fluorescence quenching effects.

- JC-1 stock solution
  To obtain a 1.0 mM solution, dissolve 1 mg of JC-1 dye in 1.5 ml of dimethylformamide. Divide unused stock solution into portions sufficient for 1 day of experimental work each and store in microfuge tubes, in a darkened container with a desiccant, at −20°C.

Equipment

- Fluorescence microscope
  This should be equipped with a suitable objective (e.g. water immersion 40× magnification, 0.75 numerical aperture).
- Optical filters
  Long-pass filter set designed for FITC, such as Omega Optical XF115 or Chroma Technology 41012. Alternatively, a filter set with separate emission windows for green and red fluorescence signals (e.g. Omega Optical XF26 or Chroma Technology 71010) can be used.
- Sterile glass coverslips
  18 mm × 18 mm square or 25 mm circular diameter)
- Sterile 35 mm petri dishes
- Glass microscope slides

Procedure

1. Culture adherent cells on coverslips inside a 35 mm Petri dish filled with the normal culture medium.

2. Prepare 10.0 μM JC-1 staining solution by adding 10 μl of 1.0 mM stock solution to 1 ml of serum-supplemented HBSS.

3. Drain the culture medium from the dish and replace with prewarmed (37 °C) staining solution (see Note 5).

4. Incubate the cells at 37 °C in a 5% CO_2 incubator for 10 min.

5. Replace the loading solution with fresh prewarmed medium. Seal the coverslip to a glass slide (e.g. with clear fingernail polish). Do not apply pressure. Examine the cells under the fluorescence microscope.

> ***Note 5*** *Solubilizing agents such as (β-cyclodextrin (50 mg/ml) may be added to the staining solution to enhance dye penetration into tissue specimens [10].*

Polarized mitochondria should be marked by punctate orange–red fluorescent staining. On depolarization, the orange–red punctate staining is replaced by diffuse green monomer fluorescence. Some of the green fluorescence may remain associated with mitochondria, due to potential-independent interactions of the JC-1 monomer with mitochondrial membranes.

3. Troubleshooting

Problem	Diagnosis
Photobleaching	Photobleaching is a deleterious process that competes with fluorescence emission, resulting in the irreversible conversion of a dye into non-fluorescent products. In microscopy, photobleaching is manifested by progressive fading of the fluorescent image during observation. Since the effect is localized to the illuminated region of the specimen, moving the field of view will typically produce a temporary restoration of the image. The most direct remedy for photobleaching is to limit the exposure of the specimen to excitation light, in terms of both duration and intensity. Antifade reagents, commonly used to inhibit photobleaching in immunofluorescence microscopy, are in most cases highly toxic to living specimens, and therefore not useful.
Dye concentrations	For reasons that were outlined in the introduction to this chapter, increasing staining concentrations is generally not recommended as a way of producing higher fluorescence signal levels. However, it is axiomatic that staining techniques that are sensitive to cell structure and function may exhibit variable performance from one cell or tissue type to another. The dye concentrations specified in Method 1 have been found to be suitable for staining NIH 3T3, PtK2 and MDCK cells, but may require adjustment for use with other cell types. Cultured mouse leucocytes (J774A.1), which have higher esterase activity levels, require 5–10 times less calcein AM, but require the same amount of EthD-1. Optimal concentrations of calcein AM and EthD-1 for assessing the viability of Schwann cells in nerve tissue preparations have been determined to be 4 and 6 μM respectively [11] and are similar to those specified in Method 1.

References

1. Mason, W.T. (ed.) (1999) Fluorescent and Luminescent Probes for Biological Activity, 2nd edn, Academic Press, San Diego, CA

2. Johnson, I. (1998) Histochem. J. **30**, 123–140

3. Bloom, J.A. and Webb, W.W. (1984) J. Histochem. Cytochem. **32**, 608–616

4. Thorpe, W.P., Toner, M., Ezzell, R.M., Tompkins, R.G. and Yarmush, M.L. (1995) Biophys. J. **68**, 2198–2206

5. Anderson, W.M. and Trgovcich-Zacok, D. (1995) Biochem. Pharmacol. **49**, 1303–1311

6. Singer, V.L., Lawlor, T.E. and Yue, S. (1999) Mutat. Res. **439**, 37–47

7. Hollo, Z., Homolya, L., Davis, W.C. and Sarkadi, B. (1994) Biochim. Biophys. Acta **1191**, 384–388

8. Smiley, S.T., Reers, M., Mottola-Hartshorn, C., Lin, M., Chen, A., Smith, T.W., Steele, G.D. and Chen, L.B. (1991) Proc. Natl. Acad. Sci. U.S.A. **88**, 3671–3675

9. Di Lisa, F., Blank, P.S., Colonna, R., Gambassi, G., Silverman, H.S., Stern, M.D. and Hansford, R.G. (1995) J. Physiol. **486**, 1–13

10. Sick, T.J. and Perez-Pinzon, M.A. (1999) Methods: a Companion to Methods in Enzymology **18**, 104–108

11. Decherchi, P., Cochard, P. and Gauthier, P. (1997) J. Neurosci. Methods **71**, 205–213

Freezing and fixation

John A. Kiernan

Chapter 6

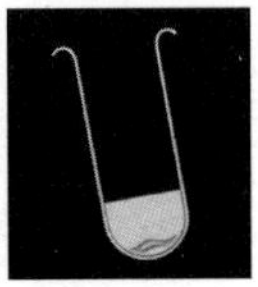

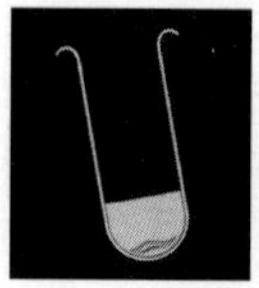

1. Introduction

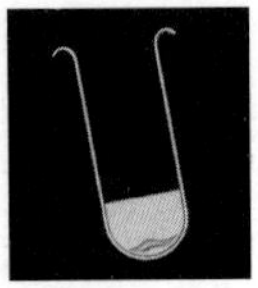

Many observations can be made on living or recently killed cells or organisms, but often it is necessary to stabilize the structural macromolecules of an object. Specimens are frequently frozen to arrest their metabolic activity or to make them hard enough to be sectioned. Freezing produces structural damage in tissues and its preservative action ceases with thawing. Lasting stabilization, usually brought about by chemical reagents, is called fixation. It has a greater effect on the eventual microscopic appearance than do any of the other events in the processing of cells or tissues. Ideally, fixation should increase the physical resilience of a specimen and render insoluble its protein, nucleic acid, lipid and larger carbohydrate molecules.

The choice of a fixative is determined by the way in which the tissue will be examined. It is, therefore, important for every investigator using histological techniques to be familiar with the effects of different fixatives on cells and tissues.

Note *You can expect difficulties and failures if you use a fixative without knowing how it works and why it might be appropriate for your research.*

Read the relevant paragraphs of explanation and discussion before making up any of the mixtures described here or elsewhere.

1.1. Reasons for fixation

There are at least five reasons for using fixed rather than unfixed material.

1. A major objective of most microscopic investigations is to obtain a picture that reveals structural and functional features of living cells.

Structure can be observed dynamically in cultures and in certain parts of whole animals or plants, but the study of a compact tissue or organ is possible only in sections thin enough to be transparent. Before cutting the sections it is necessary to stop all metabolic activity as quickly as possible.

2. Pieces of freshly removed tissue can be frozen and cut with a cryostat, and the sections will remain stable if they are quickly dried and stored below −20°C. However, the structural integrity deteriorates rapidly when the sections are exposed to an aqueous medium such as a solution of a dye or antibody. Loss of cytoplasmic components is the first and most conspicuous change. A thin preparation consisting of whole cells (such as a culture or a blood smear) is less vulnerable than a section because the cells are contained within their surface membranes.
3. Enzyme-catalysed degradation of intracellular structures (autolysis) begins soon after the death of any cell.
4. Most specimens need to be embedded in a hard matrix of wax or plastic before they can be cut into sections. The processing that precedes embedding involves passage through solvents other than water. Solvents distort the fine structure of most animal tissues and also cause extraction or artifactual displacement of substances that may need to be observed histochemically. Adequate fixation protects the cells and extracellular materials from the adverse effects of passage through solvents.
5. Changes brought about by fixation are necessary for the success of many of the techniques used to examine tissues; for example, epitopes commonly need to be exposed by distortion of the macromolecular architecture (coagulation) before they can be demonstrated immunohistochemically.

2. Freezing and heating

Freezing is not a method of fixation, but it is frequently necessary to freeze a specimen before, after, or instead of fixing it. Freezing does not denature proteolytic enzymes. Autolysis continues as soon as unfixed cells are thawed, leading to visible structural damage in both cryostat sections [1] and sections of objects that have been embedded in plastic after freeze-drying [2]. Heating a tissue coagulates proteins and is, therefore, a method of fixation.

2.1. Effects of freezing

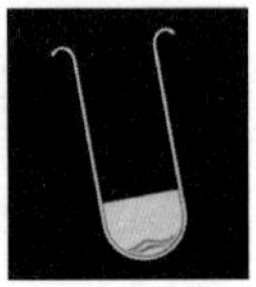

The most conspicuous changes resulting from freezing are due to the formation of ice crystals, whose positions are eventually seen as holes in the sectioned tissue. These holes may be scattered randomly through the sections or they may appear in predictable places. In skeletal muscle, for example, ice-crystal holes form in the contractile fibres rather than in the intervening connective tissue. In any specimen the number and size of ice crystals are greatest after slow freezing. Specimens frozen by simply putting them in a freezing cabinet with a temperature between −30 and −20 °C are usually useless for microscopy; cells and other components cannot be recognized in the compressed and distorted bridges of tissue between the holes in the sieve-like sections. Immersion in liquid nitrogen is almost equally destructive because even a tiny specimen becomes thermally insulated by a surrounding shell of nitrogen gas. For rapid freezing it is necessary to bring a specimen into contact with either a cold liquid that will not evaporate or a cold solid object with high thermal conductivity.

Extremely rapid freezing, achieved by slamming a thin specimen onto a metal block at the temperature of liquid helium (−268 °C) [3] or liquid nitrogen (−196 °C) [4], results in the formation of a non-crystalline (vitreous, glass-like) form of ice, but this is feasible only for tiny specimens. In larger specimens, the damage from ice crystals can be minimized (but not completely prevented) with more widely available technology. Liquid nitrogen and solid carbon dioxide (dry ice) are readily available. Ice-crystal formation can also be suppressed by infiltrating the specimen with a cryoprotective agent. Combination of rapid freezing with cryoprotection is particularly effective.

2.2. Cryoprotection

Cryoprotective agents inhibit the formation and growth of ice crystals. Most are hydrophilic organic compounds that contain numerous hydroxyl groups, e.g. glycerol, sucrose and propylene glycol. Crystallization may be inhibited by the formation of hydrogen bonds between water and the cryoprotectant molecules [5]. Naturally occurring intracellular solutes confer considerable cryoprotection, and much of the damage caused by freezing is owing to crystallization of ice between the cells. A cryoprotective agent must replace a substantial proportion of the extracellular water molecules in a specimen. Dimethyl sulphoxide [DMSO; $(CH_3)_2SO$], a cryoprotectant that can enter cells, is widely used at quite low concentrations to preserve the viability of frozen cells. Its mechanism of action is incompletely understood [6].

Cryoprotection is always desirable, and it is essential for larger specimens (greater than approx. 4 mm thick) whose central parts will otherwise be damaged by even the most rapid methods of freezing. If the requirements of the investigation allow fixation, specimens should be perfused with or immersed in an aqueous fixative mixture, such as a neutral, buffered formaldehyde solution, before immersion in the cryoprotectant solution.

Method 1. Cryoprotection with sucrose

A 60% solution of sucrose is needed to prevent ice crystal formation [7], but lower concentrations are commonly used. Any specimen may be cryoprotected with sucrose.

Materials

- Specimens
  Ideally these should have been adequately fixed, washed and equilibrated with water. The requirements of an investigation may preclude thorough

fixation, and it is then necessary to use pieces of tissue that have been exposed for only a few hours (or even minutes) to a neutral, buffered formaldehyde solution (minimal fixation). If unfixed specimens must be used they should be no larger than 1 mm in any dimension, and the times for infiltration with the sucrose solutions will be in hours rather than days.

- Cryoprotectant solution
  - *For well fixed specimens:* Add approx. 30 ml of water to 120 g of sucrose and place in a covered container in an oven at about 60 °C, and stir occasionally. When all the sucrose is dissolved (may take more than 24 h) adjust the volume with water to make 200 ml. Lower concentrations (15%, 30%) are achieved by diluting the stock 60% solution.
  - *For unfixed or minimally fixed specimens:* As above, but use 0.01 M sodium phosphate buffer instead of water. This protects against acidification due to metabolism or autolysis. Sucrose solutions are stored and used at 4 °C and must be replaced if they become cloudy from growth of micro-organisms.

Procedure

1. Place each specimen in 15% sucrose (at least ten times the volume of the specimen). Infiltration can be considered adequate when the specimen sinks. For small specimens, overnight immersion is usually sufficient.

2. Repeat the above procedure with 30% sucrose solution.

3. Ideally, move the specimen into 60% sucrose and wait for it to sink. A more common practice is to consider the 30% solution adequate.

4. Store the specimens in the highest concentration of sucrose used, until they are to be frozen.

Method 2. Cryoprotection of large specimens with glycerol and DMSO

This procedure, introduced for formaldehyde-fixed whole monkey brains [8], was found to be superior to using sucrose and some other cryoprotectant solutions.

Solution

- Glycerol
- DMSO
- Diluent (see below)
  - □ Add 20 ml of DMSO to 200 ml of glycerol and make up to 1 litre with diluent. For a well fixed specimen the diluent is the buffer solution in which the formaldehyde fixative was made up. For a minimally fixed specimen, such as an organ taken from a formaldehyde-perfused animal, the diluent is the complete fixative.

Procedure

1. Place the specimen in at least ten times its volume of the cryoprotectant solution and leave for 4 days at 4 °C.
2. Freeze the specimen by immersing it in a liquid maintained at −75 °C or colder.

   Two such procedures (using isopentane or acetone) are described later in this chapter. For large specimens the volumes will be proportionally larger than those prescribed, which are for pieces no larger than 1 cm^3.

2.3. Rapid-freezing methods

These procedures do not bring about vitrification (solidification of water to a non-crystalline type of ice), but they favour the formation of ice crystals that are too small to make holes visible by light microscopy.

The best results are obtained with fixed and subsequently cryoprotected specimens, but these procedures are also appropriate for unfixed objects.

Method 3. Immersion in isopentane at −160 °C

Materials

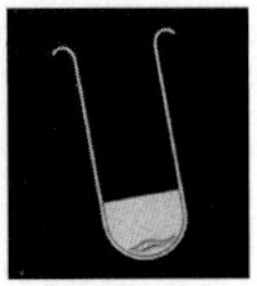

- A small metal container 2–3 cm in diameter with thin walls and a capacity of approx. 50 ml
  - □ In the past, 35 mm photographic films were sold in aluminium cans that are ideal for this purpose. Similar containers can be found in stores that sell kitchen equipment.
- A thermally insulated plastic food container approx. 10 cm wide and 10 cm deep
  - □ This too can be bought from an ordinary shop.
- Liquid nitrogen
  - □ This should be stored in a vessel that can be lifted for pouring. Allowing for evaporation, 500–1000 ml will be needed for each session. The liquid-nitrogen container can be purchased from a laboratory equipment supplier.
- Isopentane (2-methylbutane)

> ***Note*** *Isopentane (2-methylbutane) is volatile (b.p. 28 °C), highly flammable and is not miscible with water. A volume of 20–30 ml will be needed for each session.*

- Metal or plastic forceps
- Protective gloves
- Saline (0.9% NaCl in water) or an OCT (optimal cutting temperature) compound
  - □ OCT compounds are viscous hydrophilic liquids such as a poly(ethylene glycol). Several proprietary brands are available. OCT compounds provide some physical support for a specimen during sectioning, but they are dissolved by aqueous solutions used for staining. If a commercial product of uncertain composition is not

acceptable, use a 2% gelatin in saline solution. A gelatin solution (unlike OCT compounds) enters the cracks and crevices of a specimen only when warm (25–30 °C). The polymers in an OCT compound are mobile liquids at room temperature, but have much greater viscosity below approx. 5 °C.

- Airtight containers
  - Glass or plastic tubes with snap-on or lined screw-on caps are suitable. Polyethylene microcentrifuge tubes or vials for scintillation counting are cheaper than 'specimen tubes'. Each container should be of the smallest available size that will contain the specimen, and it must be kept cold (≤−20°C) until the moment it is needed (see Step 7 of the procedure below).
- Thin slices of cork (2mm)
  - These should fit easily into the airtight container.

Procedure

1. Perform the dissection or other procedure that provides the specimens. If the specimens have already been removed and fixed, cut them into small pieces. If several specimens must be frozen in one session, two people are needed, one to dissect and the other to do the freezing. A specimen should be no more than 2 mm thick. If it is to be sectioned in a cryostat, orient the piece of tissue and place it in a small drop of OCT compound or 2% gelatin on a thin slice of cork.

2. Pour liquid nitrogen into the insulated plastic container until it is two-thirds full. Half fill the metal container with isopentane.

3. Place the metal container of isopentane in the liquid nitrogen. Stir occasionally with forceps. Top up the liquid nitrogen as required to replace evaporative loss.

4. When the isopentane begins to freeze (at −160 °C a sticky, gelatinous material forms on the inside of the metal container and can be felt with forceps), place the specimen (including its cork carrier, if used) in the isopentane.

5. Wait for 30 s to allow the specimen to freeze.

6. Remove the specimen. It may first be necessary to remove the metal container from the liquid nitrogen, to melt solidified isopentane that often sticks the specimen to the side or floor.

7. Immediately, put the frozen specimen in the smallest possible airtight container, and store at a temperature no higher than −20°C.

Method 4. Contact with a cold metal surface

Materials

Some of these requirements are explained in more detail in the preceding instructions for freezing by immersion.

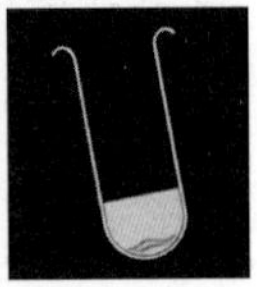

- A thermally insulated plastic food container, approx. 10 cm wide and 10 cm deep
- Liquid nitrogen
  - This should be stored in a vessel that can be lifted for pouring. Allowing for evaporation, approx. 1000 ml will be needed for each session.
- Metal or plastic forceps
- Protective gloves
- Metal microtome chuck
  - The specimen will be frozen onto this. One chuck is needed for each specimen.
- An OCT compound or 2% gelatin (as for Method 3)
- Domestic aluminium foil cut into 2 cm squares (one for each specimen)
- Airtight container to receive the frozen specimen
  - A small plastic bag is a suitable container for a metal microtome chuck with an attached specimen.

Procedure

1. Place the metal microtome chucks (one for each specimen to be frozen) in the insulated plastic container. Pour in liquid nitrogen until it is nearly full and until boiling has largely stopped. The metal chucks should now be at a temperature of −196 °C. If sections are needed through the whole

thickness of a specimen, place a drop of OCT compound or 2% gelatin (approx. 1 mm deep) onto the centre of the chuck before immersing in liquid nitrogen. This protects the blade from injurious contact with the chuck while cutting the last sections of the series.

2. Carry out the dissection or other procedure that provides the specimens. If the specimens have already been removed and fixed, cut them into small pieces. A specimen should less than 2 mm thick. If several specimens must be frozen in one session two people, are needed: one to dissect and the other to do the freezing.

3. Place the piece of tissue on the shiny side of a square of aluminium foil. The surface in contact with the foil is that from which the first sections will be cut.

4. Using gloves and forceps, lift a microtome chuck out of the liquid nitrogen and place it upon the specimen, which will adhere immediately and begin to freeze. If the chuck carries a patch of frozen OCT compound or gelatin, care must be taken in bringing it into contact with the specimen. After 5–10 s, remove the piece of aluminium foil and replace the chuck, with attached specimen, in the liquid nitrogen.

5. At the end of the session, take each microtome chuck and attached specimen out of the liquid nitrogen, put it into a suitable airtight container, and store at a temperature no higher than −20°C.

Freezing with dry ice and acetone

As an alternative to using liquid nitrogen, a cold liquid can be prepared by adding solid carbon dioxide (dry ice) to acetone in a thermally insulated container. When the dry ice stops releasing bubbles of carbon dioxide, the temperature is −78.5°C. The acetone (m.p. −94°C) does not freeze. Use thick gloves to handle dry ice, and remember that acetone is volatile and highly flammable but miscible with water.

Acetone containing dry ice can be used in the same way as liquid nitrogen or melting isopentane to freeze specimens either by immersion (in the cold acetone) or by contact with a cold surface, as described in Methods 3 and 4 respectively. Frozen specimens that are not going to be cut immediately must be stored in airtight containers at a temperature no higher than −20°C.

2.4. Freeze substitution

The purpose of freeze substitution is to preserve substances, such as amines or inorganic ions, that would normally be extracted by water. Ice is extracted from the frozen specimen by a cold solvent, which replaces the water. At the low temperature, the solvent does not denature proteins, and the tissue remains unfixed. Usually, the freeze-substituted specimen is fixed by allowing the temperature of the solvent to rise to 4 °C, and it is then processed for embedding in a hydrophobic medium, such as paraffin wax (as described below), or in a synthetic resin.

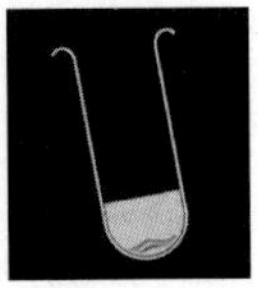

Method 5. Freeze substitution

Materials

- Specimens
  - Typically these will be fresh, unfixed objects approx. 2 mm thick.
- Arrangements for rapid freezing (see Methods 3 and 4).
- A deep-freeze cabinet that maintains a temperature of −50 °C or lower
  - True freeze substitution cannot be accomplished in a −20 °C freezer.
- Acetone
  - Approx. 20 times the volume of the specimen, in a tightly capped jar, pre-cooled in the freezer.

> ***Caution*** *Acetone is volatile and flammable*

- A clearing agent
  - Toluene or xylene are suitable, used at room temperature.

- Paraffin wax and equipment for infiltration and embedding (see Chapter 7)
- Wax-coated slides
  - These are made by mounting ribbons of 10–15 μm sections of a plain paraffin-wax block and air-drying for at least 12 h.

Procedure

1. Rapidly freeze the specimen and place it in the jar of precooled acetone.
2. Leave in the freezer for 7 days.
3. Replace the cold acetone and leave in the freezer for another 7 days.
4. Move the jar into a refrigerator (approx. 4 °C) and leave it there for approx. 2 h. The solvent, warmed to 4 °C, coagulates proteins.
5. Replace the acetone with a clearing agent (xylene or toluene). Leave for 2 h at room temperature (see Chapter 7).
6. Transfer the specimen to a second change of the clearing agent for 2 h, then infiltrate with paraffin wax (three changes, each 2–3 h, at 56–58 °C) and embed in the usual way.
7. Mount the sections on dry slides without floating on water, as this would extract the substances preserved by freeze substitution. Such sections can be difficult to flatten, and they tend not to adhere well to the slides. To facilitate flattening, place the sections on wax-coated slides and manipulate with needles on a hotplate maintained at approx. 60 °C. Unfortunately, this procedure increases the likelihood of detachment, because there is a layer of wax between the sections and the glass.

2.5. Fixation by heat

Most of the protein in a specimen can be coagulated (made permanently insoluble without chemical alteration of the amino acids) if the temperature is raised to 55 °C. Bacterial smears are commonly fixed by stronger heating (holding the slide over a flame) because the shapes and some of the staining properties of these cells are preserved by this simple, but extreme, procedure. Chemical fixation is mandatory for the examination of structures within bacteria, such as nucleoids [9].

2.5.1. Microwave heating

Solid objects must be heated evenly throughout, to prevent considerable distortion of the architecture of the tissue. This is achieved by immersing the specimen in approx. 30 ml of saline and heating in a microwave oven. Microwaves agitate polar molecules such as water and fat; the increased movement constitutes a rise in temperature.

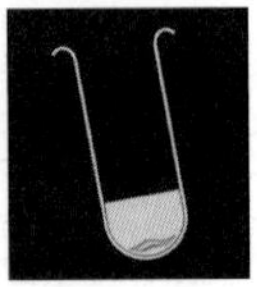

Specialized microwave ovens for laboratory use are more reliable than domestic models, but are extremely expensive. Domestic microwave ovens are suitable, but must first be calibrated to determine the optimum position for placement of the specimen (or slides), to ensure that the desired rise in temperature will be obtained in 20–60 s [10,11]. Thermometers and temperature-sensitive tapes suitable for use in microwave ovens are sold by scientific supply companies. The magnetron (microwave generator) does not begin to emit microwaves until the oven has been switched on for 5–10 s, and the emission is not continuous but in pulses of 2–10 s. Ovens should be used at full power because at lower settings ('defrost' etc.) the pulses are more widely spaced and the timing of the irradiation is less precise.

The specimen container, a glass or plastic vessel, must be square or rectangular in section, because total internal reflection of microwaves at the interface of water with the walls of the vessel results in the concentration of the heating effect of the radiation in the centre of a 3–4 cm diameter cylinder. The oven cavity should also contain approx. 200 ml of water in a glass or plastic jar. This 'water load' protects the magnetron by absorbing the radiation that is reflected from the walls of the cavity but is not absorbed by the specimen and its surrounding fluid. The distribution of radiation within the cavity of a microwave oven is not uniform. When suitable positions have been determined for the specimen container and water load, these are marked on the floor of the oven. If there is a rotating table, the optimum starting postion should be marked on the underlying floor. This may not be the centre of rotation.

For any individual microwave oven the marked positions and the volumes of the specimen container and 'water load' should always be the same. The duration of microwave irradiation will vary with the medium surrounding the specimen. The optimum time must be determined for each liquid. Solutes such as salts or organic compounds generally enhance the absorption of microwaves by water, shortening the time needed to achieve a temperature of 55 °C.

2.5.2. Effects and uses of microwave heating

Artifacts, notably small distorted nuclei and coarsely textured cytoplasm, are conspicuous in sections of tissues that have been fixed solely by heating, especially if the temperature becomes too high (e.g. 80 °C) [12]. For this reason fixation of solid specimens by heat alone is rarely satisfactory. A much more important application of microwave heating is to accelerate chemical fixation. If specimens are 'microwaved' to 55 °C in a formaldehyde-containing solution, the penetration and the chemical reactions of this fixative are greatly accelerated. Fixation is partly chemical and partly by thermal coagulation, and in a few minutes it is possible to achieve effects similar to those obtained by several hours of immersion in the same fixative at room temperature.

In addition to fixation, microwave heating can be used to accelerate many other activities in the histological laboratory. These include penetration of dehydrating solvents and staining of sections mounted on slides. Detailed instructions are given by Kok and Boon [10].

When moderate heat must be delivered with precision to a small object, such as a single section covered by a tiny drop of an antibody solution, it is necessary to use an expensive microwave oven designed for laboratory use.

3. Principles of chemical fixation

3.1. Chemical coagulation

Several liquid reagents can render proteins insoluble by disturbing the structures of the polypeptide chains. Those used in histological fixatives bring about a coagulation that is less coarse than that induced by heat, but nevertheless often visible by light microscopy. Organic solvents, such as ethanol or acetone, are often applied to thin layers of smeared or cultured cells and to sections of previously unfixed specimens that have been frozen, cut with a cryostat and mounted onto slides or coverslips. In freeze substitution, an organic solvent at a low temperature (below −49 °C) extracts ice from a frozen specimen without coagulating proteins. Coagulation occurs uniformly throughout the dehydrated specimen when the temperature is allowed to rise. The use of freeze substitution retains water-soluble substances, e.g. calcium and many other inorganic ions, that do not dissolve in organic solvents. For other purposes, notably fixation of blocks of tissue, coagulants are almost always components of mixtures with other reagents.

Organic solvents, e.g. the lower alcohols (methyl, ethyl, isopropyl) and acetone, are examples of non-aqueous coagulant fixative agents. The coagulants commonly used in aqueous solution are picric acid, mercuric chloride ($HgCl_2$) and salts of some other metal ions, notably zinc (Zn^{2+}).

Picric acid (trinitrophenol) precipitates proteins by forming salt linkages with amino groups and other basic groups, and may also be bound partly by weaker forces known as charge-transfer bonds [13]. It causes visible coagulation of nucleoplasm and cytoplasm, with distortion of organelles. There is shrinkage of the whole specimen, but not hardening [14]. These properties facilitate infiltration with embedding media and cutting of sections.

Caution *Picric acid explodes when heated to 300 °C, which is a theoretical possibility if undue force is applied to a dried-on bottle cap or glass stopper. It is always stored under water, as a yellow sludge. A saturated aqueous solution (1.3%, pH 1.5–2.0) is used for making up fixatives and stains. A saturated alcoholic solution (about 8.5% in 80–95% ethanol) is used for making non-aqueous mixtures.*

Mercuric chloride also combines with amino groups, with which it forms weak covalent bonds. In addition, strong covalent bonds are formed between mercury atoms and the sulphydryl side-chain of the amino acid cysteine. Mercuric chloride brings about a finer (less lumpy) cytoplasmic insolubilization than any other coagulant fixative. The coagula, which are 40–200 nm wide [15], are mostly too small to be resolved by light microscopy, thereby resulting in the preservation of most organelles and secretory granules. Tissues fixed in mercuric chloride contain evenly distributed deposits of a granular, dark insoluble substance, probably mercurous chloride (Hg_2Cl_2). Surprisingly, this 'mercury pigment' does not cause structural disruption, but it must be removed. This is achieved by treatment with iodine, applied either to the whole block before embedding or to sections before staining.

Caution *Mercury-containing compounds are dangerous poisons and must not be discarded into public drains. In several modern mixtures, mercuric chloride is replaced by zinc chloride or zinc sulphate, which also coagulate proteins and are less toxic. Many workers consider the quality of nuclear and cytoplasmic fixation by zinc to be somewhat inferior to that attainable with mercury.*

Acetic acid is a component of many fixative mixtures, at concentrations of between 5% and 33% (v/v). It does not render proteins soluble, but it extracts some nucleohistones, the proteins that support the DNA molecules in the chromosomes, and it also brings about a delicate precipitation of DNA [14]. Consequently, in a tissue fixed in the presence of acetic acid the cells acquire characteristic patterns of nuclear chromatin. These appearances may be largely artifactual, but they are useful for the recognition of different cell types in stained sections. If used alone, acetic acid can cause swelling of tissues (especially those containing collagen), with much structural disruption, but this adverse effect is largely opposed by other components of a fixative mixture, such as ethanol or picric acid, which tend to cause shrinkage.

3.2. Aldehyde fixatives

The most widely used cross-linking fixatives are aldehydes. These combine with the amino and amide (peptide) nitrogen atoms of proteins, causing covalent cross-linking if these nitrogen atoms are suitably positioned [5,16]. There are other cross-linking agents, notably the dichromate ion (at $pH > 4$) and various carbodiimides, that form covalent bridges between carboxyl groups. Extensive cross-linking within and among cells confers considerable physical resilience (similar methods are used for changing dermal collagen into leather), and preserves the structural details that coagulation would destroy. Cross-linking of protein also traps certain water-soluble substances within the tissue; these substances, such as mRNA, various large carbohydrate molecules and some lipids, do not react with the fixative. A serious detrimental consequence of cross-linking is steric hindrance of diffusion of large molecules. Embedding media, especially melted wax, may not adequately infiltrate the specimen. Antibodies, enzymes and other proteins used as reagents may fail to reach their specific targets, even when applied to thin sections. Immunostaining is also prevented when chemical reaction with the fixative renders an epitope unrecognizable to its antibody. Some cross-linking reactions, notably those of formaldehyde, can be reversed with a variety of antigen retrieval methods.

Formaldehyde is a gas that dissolves in and reacts with water. It is most commonly obtained as formalin, which is a 37–40% solution by weight, to which about 10–15% methanol is added by the manufacturer to hinder polymerization. Most of the formaldehyde in formalin is present in the form of low polymers [16]. These break up into monomeric formaldehyde molecules when an excess (typically 10 volumes) of water is added; the depolymerization is accelerated by addition of acid or base or by buffering to an approximately neutral pH. An addition reaction of formaldehyde monomer with water produces methylene hydrate, $CH_2(OH)_2$, which is the chemically reactive component of the fixing solution.

An alternative source of formaldehyde is a solid polymer, paraformaldehyde. This is a white powder that can be added to a suitable aqueous buffer solution, and then depolymerized ('dissolved') by heating to 60 °C. The buffer salts are necessary for the depolymerization to occur. The absence of 1–1.5% methanol from a 3.7–4.0% solution prepared from paraformaldehyde is the only difference between it and a 10-fold dilution of formalin [17]. The composition of any aqueous formaldehyde solution changes over a long storage time because of the Cannizzaro reaction, in which formaldehyde molecules react with one another to generate methanol and formic acid. Eventually, the concentration of formic acid may make the fixative too acidic for its intended purpose, but this is unlikely to occur in neutral, buffered formaldehyde solutions, even after storage for several months.

The small molecules of methylene hydrate penetrate tissues rapidly, but their reaction with protein occurs slowly. It is, therefore, important to prevent disruption of cells before they are chemically fixed; the formaldehyde must be dissolved in a liquid whose osmotic pressure and pH are similar to those of the extracellular fluids of the specimen being fixed. For mammalian organs simple saline (0.9% NaCl; 0.153 M) is sometimes used, but a 0.03–0.05 M sodium phosphate buffer (pH 7.2–7.4) is preferable. Calcium acetate (2%; 0.04 M) is an alternative, and the calcium ions contribute to the retention of phospholipids in cell membranes [18,19]. Sea water is often used as the diluent when marine specimens are to be fixed.

Fixation by formaldehyde is attributable to covalent cross-linking of protein molecules by short methylene bridges. Non-protein materials such as nucleic acids and carbohydrates do not react but are entrapped in the cross-linked proteinaceous matrix. The nuclei of cells that have been fixed only in formaldehyde do not show well developed chromatin patterns. When formaldehyde is the only chemically reactive component of a fixative mixture, structural stabilization of the tissue requires immersion for 1–2 weeks. Much shorter times are used in practice, and the incompletely fixed specimens remain osmotically responsive [20]. Proteins in such specimens may be extracted by aqueous reagents or coagulated by liquids such as ethanol. The components of a formaldehyde-fixed tissue shrink differently during later processing, and artifactual spaces around cells and blood vessels are frequently seen, especially in paraffin sections.

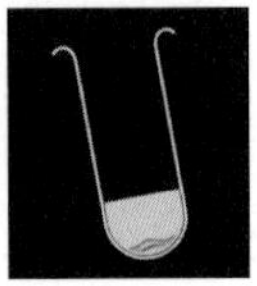

Glutaraldehyde is the other cross-linking fixative in common use. It is supplied as a concentrated (25 or 50%) aqueous solution, which is diluted to 0.5–2.5% for use. Glutaraldehyde is a dialdehyde, but its solutions contain polymers of various sizes, each with two or more free aldehyde groups [21]. These react principally with nitrogen atoms of proteins, allowing the formation of cross-links of variable length. Aged or 'technical grade' glutaraldehyde is unsuitable for use as a histological fixative because it consists largely of polymers too large to enter the proteinaceous tissue matrix. A reagent of high quality ('electron microscopy grade') must be used. Glutaraldehyde penetrates much more slowly than formaldehyde, but the chemical reactions with protein occur almost instantaneously.

Glutaraldehyde was introduced [22] as a replacement for osmium tetroxide (OsO_4), which previously was the only fixative agent that provided understandable structural preservation for electron microscopy. It soon became evident that the penetrating power of formaldehyde could be combined with rapid and extensive chemical cross-linking by glutaraldehyde. Mixtures of these two aldehydes have been used routinely as fixatives for electron microscopy since 1965. Osmium tetroxide is still used as a postfixation to impart additional electron opacity to fixed specimens, but not as the primary fixative.

The extensive cross-linking brought about by glutaraldehyde provides more life-like ultrastructural preservation than any other fixative, but at the expense of considerable chemical change. Many enzymic activities are suppressed, and antigenic determinants are blocked by combination with the fixative. Glutaraldehyde-fixed material also contains abundant unreacted aldehyde groups; these combine non-specifically with antibodies and other proteinaceous reagents, and they also interfere with many traditional staining methods. The free aldehyde groups must be reduced (by sodium borohydride) or blocked (by reaction with phenylhydrazine or certain other compounds) before attempting to apply any immunohistochemical procedure to sections of glutaraldehyde-fixed tissue [23]. Another consequence of extensive cross-linking is impairment of the penetration of large molecules of an embedding medium such as paraffin wax. If a glutaraldehyde-fixed specimen must be embedded in wax, its smallest dimension should no greater than 1 mm. Larger fixed specimens can be frozen or processed for plastic (resin) embedding. Methacrylate, epoxy- and other monomers are small enough to enter an extensively cross-linked tissue.

3.3. Other fixative agents

Traditional cytological fixatives contain osmium tetroxide and potassium dichromate. These oxidizing agents react with and render insoluble lipids and proteins. Tissues fixed in glutaraldehyde and/or formaldehyde are commonly post-fixed in osmium tetroxide to confer electron opacity, which results principally from the binding of reduced osmium compounds to membranes. Acidified dichromate solutions (chromic acid) preserve chromosomes and the mitotic spindle well, but cause severe cytoplasmic damage. At neutral pH, the dichromate ion cross-links cytoplasmic and extracellular proteins via their carboxyl groups, and is used in micro-anatomical and cytoplasmic fixatives. The cross-linking is probably done mainly by chromic ions (Cr^{3+}) formed by local reduction of dichromate [24,25]. Prolonged treatment of a specimen with neutral dichromate renders phospholipids insoluble, enabling them to be stained in sections even after dehydration and paraffin embedding [26,27].

4. Fixation for immunohistochemistry

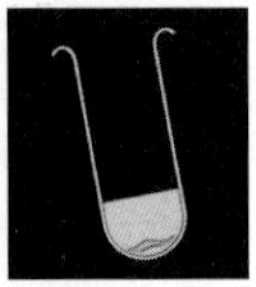

Coagulant ingredients of a fixative often expose antigenic determinants, thereby facilitating immunohistochemical staining, but some antigens are destroyed by such treatment and can be detected only in unfixed or minimally fixed material. Despite its shortcomings, neutral formaldehyde is the fixative most frequently used in research, especially when frozen or cryostat sections are to be cut. There are several ways to treat sections of formaldehyde-fixed tissue with hot water to hydrolyse the methylene bridges, thereby undoing some of the chemical cross-linking and making antigenic determinants accessible to antibodies. These procedures, collectively known as antigen retrieval, are outside the scope of this chapter.

5. Fixative procedures for general purposes

The procedures and mixtures that follow are suitable for immersion of small pieces of animals or plants, ideally with the smallest dimension being 5 mm or less (1–2 mm for slowly diffusing fixatives, notably glutaraldehyde and osmium tetroxide).

5.1. Fixation by immersion

The delay between removal of the specimen and contact with the fixative should be as brief as possible, in order to minimize *post mortem* changes, especially autolysis. Injection of fixative solution into a tube or cavity, such as an artery, the trachea or the lumen of the digestive tract, accelerates the inhibition of autolysis. With deeply anaesthetized laboratory animals, perfusion through the vascular system brings the fixative rapidly into contact with all parts of the body.

Sometimes, much larger objects, notably brains of already dead mammals, must be fixed by immersion, because trimming the soft, unfixed organ would cause extremely severe damage. During a delay of several hours before contact with an inwardly diffusing fixative, the changes that occur in the deeper regions are much less conspicuous than the effects of attempting to make clean cuts through a large, unfixed, soft specimen. Vascular perfusion of a removed organ is possible, although troublesome, and the flow of fixative is impeded by clotted blood in some of the vessels.

Care must be taken not to damage specimens during dissection. Severe abnormalities can be seen, for example after indenting the surface of central nervous tissue with a hard instrument or a fragment of bone [28,29]. The diagnostic value of a lymph node can be lost if its connective-tissue capsule is squeezed by a ligature or haemostat applied during surgical removal. Small objects may be immersed whole. Organs too large to be fixed whole must be cut carefully, usually with a sawing motion, using an extremely sharp edge. A new razor blade or disposable scalpel blade should be used for no more than ten cuts, even through soft material. Tissues that contain much water, including the softer parts of plants and many invertebrates, are severely damaged near the sites of cutting. Such specimens should be trimmed again after fixation (for example, when in 70% ethanol) to remove the damaged 1–2 mm next to the original cut surface.

5.2. Fixation by perfusion

When a fixative is circulated in the vascular system, the distance for diffusion can be reduced to that separating a cell from its nearest capillary.

The following procedure is applicable to rodents such as rats. For larger or smaller animals the volume of fixative needs adjustment — an approximate rule is that the volume in millilitres should be approximately the weight of the animal in grams.

Method 6. Fixation by perfusion

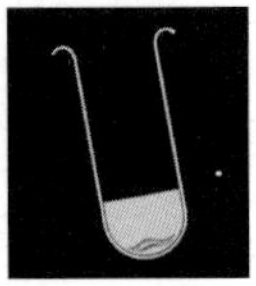

Materials

- A solution of an injectable anaesthetic agent such as pentobarbital sodium (60 mg/ml), chloral hydrate (300 mg/ml) or ketamine hydrochloride (10 mg/ml)
  - The last is a 10-fold dilution of the veterinary product.
- A saline solution
  - Aqueous sodium chloride (9 g/1000 ml NaCl) can be used. Alternatively, sodium nitrite ($NaNO_2$, 1 g/1000 ml), which dilates blood vessels, and/or heparin (100 units/1000 ml), which is an anticoagulant, can be used.
- Fixative (1 litre)
  - Usually this is a neutral formaldehyde solution or a mixture containing formaldehyde and glutaraldehyde.
- An arrangement for perfusion such as that shown in Fig. 1
- Scalpel, scissors and stout forceps

Procedure

1. Place approx. 1000 ml of saline in one bottle and approx. 1000 ml of fixative in the other. Run both liquids through the tubing to expel air bubbles, then close the tap from the fixative and allow the tubing to fill with saline. Attach a hollow needle or cannula (internal diameter $\geq$1 mm) to the end of the tube.

2. Inject the animal intraperitoneally with one of the anaesthetic solutions described above (0.2 ml/100 g body weight). (This is more than double the dose that would be used for surgery.)

3. When the animal is deeply asleep, make a midline incision in the skin of the thorax and upper abdomen. Pull apart the flaps of skin to expose the ribs and muscular abdominal wall. Using scissors, cut through the ribs on each side and join the cuts by snipping horizontally through the upper end of the sternum. Remove the anterior thoracic wall by cutting through the ventral edge of the diaphragm. The heart is now exposed (lower part of Fig. 1).

4. Hold the apex of the heart with forceps and insert the needle or cannula into the left ventricle. The depth of penetration should be less than 5 mm. While still holding, with one hand, the forceps and the tube leading into the cannula, use scissors in the other hand to make a snip in the right atrium. Open the tap to allow saline to flow into the heart.

5. Release the forceps but retain a hold on the tubing. The thoracic cavity quickly fills with blood, which overflows into the sink. Allow the saline to flow until the fluid coming out of the atrial incision is clear. This usually takes approx. 1 min. If the rate of flow seems inadequate, gently shift the position of the end of the cannula, but do not pull it out of the heart.

6. Turn off the flow of saline and turn on the tap for the fixative. The animal's muscles stiffen during the first 30–60 s of circulation of the fixative. Do not let go of the tubing until all movement has ceased.

7. When 150–200 ml of fixative has perfused (for a 300 g rat this takes 5–10 min), turn off the fixative tap.

8. Subsequent actions depend on the requirements of the investigation.

   - *Minimal fixation*. Run 150–200 ml of saline through to wash out the fixative, remove the cannula from the heart, dissect out the specimens and put them in a cryoprotectant solution as described earlier in this chapter (Section 2.2).
   - *Routine fixation*. Remove the cannula from the heart, dissect out the specimens and put them in fixative solution for the required length of time, as explained earlier in the section on aldehyde fixatives (Section 3.2) and also in the next section of this chapter.

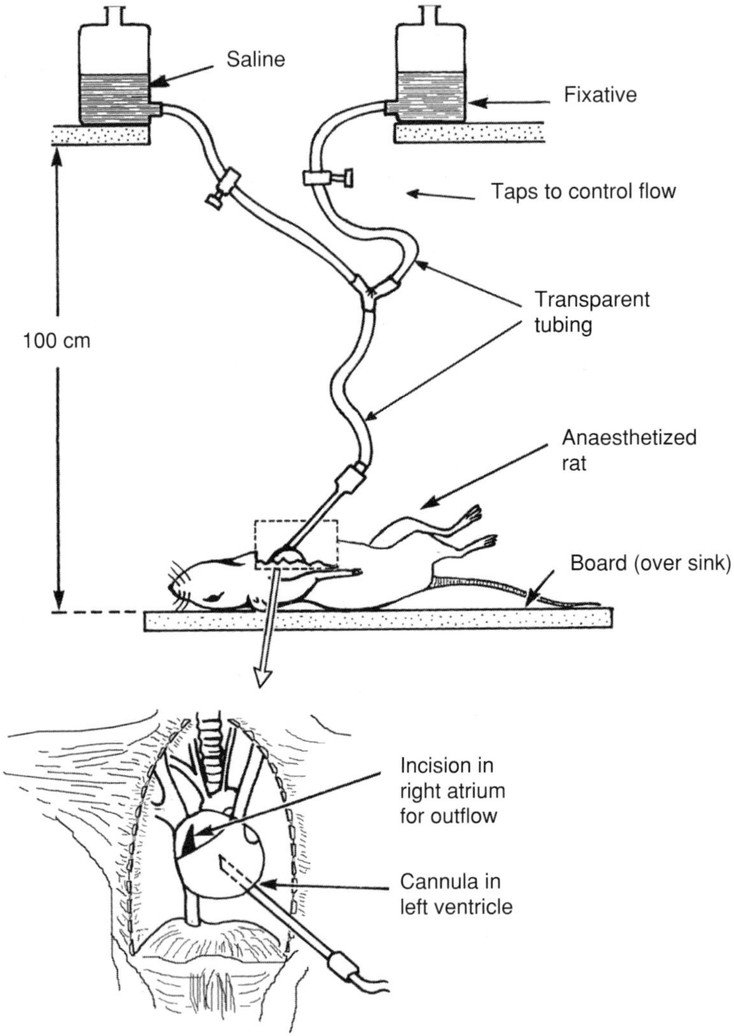

Fig. 1. Perfusion of a small animal with saline, followed by a fixative

6. Fixative mixtures

Different chemical fixative agents are mixed with one another and with other substances, in the hope of combining the advantages of all ingredients. Mixtures of coagulant and cross-linking agents generally provide optimal structural preservation for subsequent paraffin embedding and light microscopy, but at the expense of some chemical changes, including the loss of most enzymic activities. When formaldehyde is mixed with one or more coagulants, the chemical cross-linking of protein molecules occurs after the rapid changes of coagulation. Acetic acid improves the appearance of nuclei in light microscopy, and some mixtures contain chemically unreactive ingredients such as buffers and sodium chloride or sodium sulphate, to provide pH and osmotic pressure similar to those of extracellular fluids.

6.1. Formaldehyde fixatives

There are several ways to make an approximately isotonic formaldehyde solution, buffered to pH 7.2–7.4; three are given here. The solutions can be stored for several months, either at room temperature (15–25 °C) or in a refrigerator (4 °C). The concentration of formaldehyde is not critical. The usual concentration is 3.7–4.0%, but 1% is also effective. It has been claimed that immersion in unbuffered, undiluted formalin (40%, present largely as low polymers) provides adequate fixation of animal brains for subsequent cutting of thick frozen sections.

6.1.1. Phosphate-buffered formalin [30]

- 100 ml of formalin (37–40% formaldehyde)
- 900 ml of water
- 18.6 g of sodium phosphate ($NaH_2PO_4 \cdot H_2O$); monobasic
- 4.2 g of NaOH

6.1.2. Formal-calcium acetate [18]

- 100 ml of formalin (37–40% formaldehyde)
- 20 g of calcium acetate [$(CH_3COO)_2Ca \cdot H_2O$]
- Water (to make volume up to 1000 ml)

6.1.3. Formaldehyde from paraformaldehyde

Add the following in a 2 litre beaker or flask on a heated magnetic stirrer in a fume cupboard:

- 4.0 g of sodium phosphate, monobasic ($NaH_2PO_4 \cdot H_2O$)
- 4.0 g of sodium phosphate, dibasic (Na_2HPO_4)
- 40 g of paraformaldehyde
- 900 ml of water

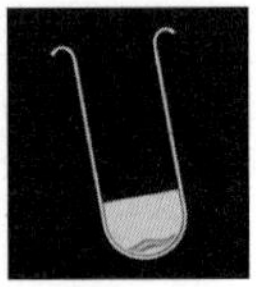

Heat with vigorous stirring until the temperature reaches 60 °C, then turn off the heat. Continue stirring (usually for approx. 5 min) until the solution is transparent. Allow the mixture to cool, add water to make the volume up to 1000 ml, and transfer to a screw-cap bottle for storage.

6.1.4. Formaldehyde with glutaraldehyde

Glutaraldehyde is included in fixatives for electron microscopy and is occasionally used for high resolution studies by light microscopy. Its limitations in relation to paraffin embedding and histochemistry have already been mentioned. It is the usual practice, although certainly not essential, to start with a formaldehyde solution made from paraformaldehyde. Concentrated glutaraldehyde solution should be added on the day the fixative is to be used.

A glutaraldehyde concentration in the range 0.5–2.5% is obtained by adding enough 25% or 50% glutaraldehyde to 90–100 ml of a buffered formaldehyde solution. The original mixture, described in an abstract by Karnowsky [31], contained 2.5% glutaraldehyde. Lower concentrations are much more popular, and are often identified by informal names such as 'half-strength Karnowsky'.

Although the concentrations of formaldehyde and glutaraldehyde are not critical, the concentrated solution of glutaraldehyde must be one designated by a reputable supplier as being suitable for electron microscopy. It must also be used before its expiry date. Old or cheap glutaraldehyde solutions consist largely of polymers too large to penetrate the tissue properly, even after vascular perfusion.

6.1.5. Procedures for minimal formaldehyde fixation

Any neutral formaldehyde solution may be used. For unfixed cryostat sections collected onto slides or coverslips, immerse in the fixative for approx. 1 min, then rinse in saline (0.9% NaCl). For solid specimens to be taken from a small animal, perfuse the fixative from the left ventricle to the right atrium for approx. 5 min, then perfuse saline for 5 min. Dissect out the desired specimens and place them in 15% sucrose at 4 °C. Improved cryoprotection can be achieved with higher concentrations of sucrose, as described elsewhere in this chapter (Section 2.2). The specimens are then frozen and sections are cut with a freezing microtome or a cryostat. If thick sections (50–200 μm) can be used, they can be cut with a vibrating microtome, thereby avoiding the need for cryoprotection or freezing.

6.1.6. Procedures for adequate formaldehyde fixation

Any neutral formaldehyde solution may be used. For solid specimens to be taken from a small animal, perfuse the fixative from the left ventricle to the right atrium for approx. 5 min, then dissect out the desired specimens. Immerse them in the same fixative at least overnight, but preferably for approx. 1 week. The temperature is not critical. If vascular perfusion is not feasible, cut the specimens into pieces no greater than 5 mm thick before immersing them in the fixative.

For subsequent enzyme histochemistry (to detect esterases and other enzymes that are resistant to formaldehyde), it is usual to fix overnight at 4 °C, and then to cryoprotect with sucrose and to cut frozen or cryostat sections. For subsequent paraffin or plastic embedding, wash the well fixed specimens in water for a few hours (to remove salts that are insoluble in ethanol), then proceed with dehydration, clearing, infiltration and embedding, as described in Chapter 7.

6.2. Micro-anatomical fixatives

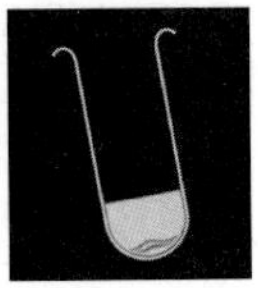

These mixtures all contain at least one coagulant, often mixed with formaldehyde to provide some cross-linking and acetic acid to coagulate nuclear DNA. Compared with formaldehyde alone, these mixtures provide fewer artefacts of differential shrinkage, more informative structural details within nuclei, and brighter staining by many methods, especially those using two or more anionic dyes in combination. Erythrocytes may have distorted shapes and stain weakly and unevenly after fixation in aqueous micro-anatomical fixatives. Their delicate membranes are damaged and made permeable by contact with the non-isotonic solution; water enters, and haemoglobin escapes before being rendered insoluble. With isotonic formaldehyde, and also with largely non-aqueous fixatives, the haemoglobin is immobilized *in situ*, and the shapes of the erythrocytes are also well preserved.

6.2.1. Modified Carnoy's fixative [32]

This non-aqueous mixture coagulates proteins and DNA, and extracts most lipids, including those of membranous organelles, but it also provides excellent structural preservation in paraffin sections. The low concentration of acetic acid (5% rather than the 10% of traditional Carnoy) prevents extraction of RNA.

- 60 ml of ethanol (100%)
- 30 ml of chloroform
- 5 ml of glacial acetic acid

Prepare immediately before using. Fix specimens for 6–8 h, then transfer them to 95 or 100% ethanol.

6.2.2. Davidson's fixative [33]

This is one of many alcohol/formalin/acetic acid mixtures that contain ethanol as a protein coagulant, formaldehyde to cross-link proteins and acetic acid to precipitate nuclear chromatin. It is used to fix cells that have been smeared onto slides. Some pathologists find Davidson's fixative useful for large, fatty specimens.

- 40 ml of 37% formalin
- 60 ml of 95% ethanol
- 20 ml of glacial acetic acid
- 60 ml of tap water

The mixture can be stored for approx. 1 month. Cells are fixed by immersion of the slides for 30 min. Solid objects are immersed for 12–24 h, and then transferred to 70% ethanol.

6.2.3. Bouin's fluid

This is a mixture of coagulant (picric acid) and cross-linking (formaldehyde) ingredients, with acetic acid to precipitate DNA and constrain shrinkage. Delicate cytoplasmic organelles, such as mitochondria, are not preserved, and acid hydrolysis and extraction of cytoplasmic RNA can occur. Prolonged immersion in Bouin's fluid also causes partial hydrolysis of DNA, which then becomes stainable with reagents that detect aldehyde groups (the Feulgen reaction). These chemical changes are avoided by fixing for less than 24 h.

- 750 ml of saturated aqueous picric acid solution
- 250 ml of 37–40% formalin
- 50 ml of glacial acetic acid

This solution can be stored for at least 5 years. Specimens are fixed by immersion for 12–24 h at room temperature and are then transferred into 70% ethanol.

6.2.4. Gendre's fluid ('alcoholic Bouin')

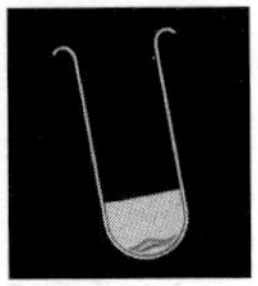

This non-aqueous mixture is useful for preserving glycogen (which is at least partly extracted and displaced within the tissue by most aqueous fixatives) and some other carbohydrates. It is also used for insect and other invertebrate tissues. Saturated alcoholic picric acid is made by adding approx. 17 g of wet picric acid powder to 200 ml of 95% ethanol.

- 200 ml of saturated alcoholic picric acid
- 38 ml of 37–40% formalin
- 12 ml of glacial acetic acid

This mixture can be stored for several years. Its chemical composition changes with time, providing improved fixative properties for some tissues [34]. Specimens are fixed by immersion for 12–24 h and are then transferred to 70% ethanol.

6.2.5. Zinc/formalin mixtures

Many mixtures of zinc salts (coagulants) with formaldehyde (cross-linker) are in current use, especially for subsequent immunohistochemical staining of paraffin sections. It is held by some that the zinc salts provide a satisfactory, but less toxic, substitute for mercuric chloride. One of the first formaldehyde-containing mixtures to be used as a fixative was introduced in 1895 by Fish [35].

- 2000 ml of water
- 50 ml of formalin
- 100 g of NaCl
- 15 g of $ZnCl_2$

Fish fixed large specimens (whole human brains) by immersion for 7–10 days. More recent mixtures contain more formaldehyde, less zinc and less NaCl than Fish's [23,36] and provide better structural preservation. A simple mixture recommended by Churukian [37] is:

- 100 ml of 37% formalin
- 1000 ml of water
- 10 g of $ZnCl_2$

Smaller specimens (2 mm thick) are fixed well enough for routine histopathology by immersion in zinc/formalin overnight, but they may remain in the solution for 1 week to allow the formaldehyde to cross-link proteins. Specimens must be washed in water to remove salts before cryoprotecting, freezing or processing in paraffin.

6.2.6. Stieve's fixative

This is one of several 'formal-sublimate' mixtures that contain mercuric chloride and formaldehyde. Such mixtures provide excellent preservation of cytoplasmic and nuclear detail and are notable for the brilliant staining obtained with dyes applied to paraffin sections.

- 20 ml of 37–40% formalin
- 5.6 g of mercuric chloride
- 4 ml of glacial acetic acid
- Water (to make the volume up to 100 ml)

The solution can be stored for several years. Fix by immersion for 12–24 h, then transfer the specimens to 70% ethanol. To remove mercury deposits, either store the specimen for 12 h in 0.5% iodine (in 70% ethanol), or treat hydrated sections with 0.5% iodine (in 70% ethanol) for 5 min, and then transfer to 2% aqueous sodium thiosulphate ($Na_2S_2O_3 \cdot 5H_2O$) for approx. 30 s to remove the yellow iodine stain, and wash for 2 min in running tap water.

Caution *Mercury-containing solutions must not be discarded by pouring down drains. Add thioacetamide (2 ml for each 100 ml of fixative) and wait 24 h for the mercury to be precipitated as black mercuric sulphide. Collect the precipitate by filtration, and keep it for later safe disposal in compliance with local laws and environmental regulations.*

6.3. Fixatives for chromosomes

Methods for counting chromosomes and examining their banding patterns generally involve arresting cell division at metaphase, mechanical elimination of most extracellular and cytoplasmic structures, and spreading of the mitotic figures on the surface of a glass slide. The starting material is either a fragment of tissue that is rich in mitotic or meiotic cells (such as a root tip, an anther, or seminiferous tubules) or a cell culture (such as leucocytes that have been stimulated to divide by treatment with a lectin). Fixation is commonly accomplished by brief exposure to a mixture of ethanol with acetic acid. In some procedures the stain is dissolved in the fixative. Detailed instructions for these techniques are given by Presnell and Schreibman [36] and Clark [38].

Dividing cells in sections have traditionally been fixed by solutions that contain chromium trioxide (CrO_3), osmium tetroxide (OsO_4) and, often, acetic acid. Chromium trioxide (dichromate at pH <3.5) preserves the outlines of cells and nuclei and coagulates the chromatin, thereby enhancing the visibility of the condensed chromosomes of dividing cells.

The mitotic spindle is also preserved and has a fibrous appearance [39]. The cytoplasm is coagulated and organelles cannot be seen in it. A typical mixture of this kind is the 'strong fluid' of Walther Flemming, who described the events of mitosis animal cells in 1882.

6.3.1. Flemming's strong fluid

This solution is made when required from stock solutions that can be stored for several months. The osmium tetroxide stock solution must be kept in a completely airtight bottle (to avoid contact with oxidizable substances such as dirt, grease or plastic). A black rubber bung contains OsO_4 vapour more effectively than a ground glass stopper.

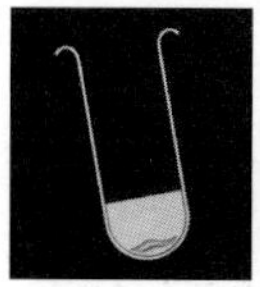

- 2.0 ml of 2% aqueous osmium tetroxide
- 1.5 ml of 5% aqueous chromium trioxide
- 0.5 ml of glacial acetic acid
- Water (to make volume up to 10 ml)

Immerse tiny fragments (1 mm or smaller) for 6–18 h. Wash in water for 6–8 h, changing hourly until no more orange/yellow colour (dichromate ions) is extracted from the specimen.

> ***Caution*** *The vapour from osmium tetroxide can be injurious to the eyes and respiratory tract.*

> ***Disposal of used fixative*** *Add an equal volume of ethanol or methanol and leave overnight. This reduces the* CrO_3 *and osmium tetroxide to insoluble, non-volatile and relatively unreactive compounds.*

6.4. Fixatives for cytoplasmic structures

Stabilization of organelles is achieved by exposure to osmium tetroxide and either mercuric chloride or potassium dichromate (pH > 3.5) or both. A typical cytoplasmic fixative for light microscopy is Altmann's mixture.

6.4.1. Altmann's fixative

This solution is made when required from stock solutions. See the above notes on Flemming's strong fluid for stability and safety precautions.

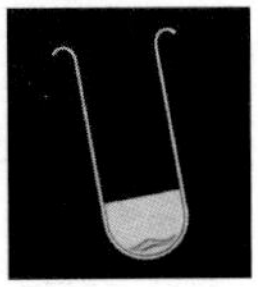

- 5.0 ml of 2% aqueous osmium tetroxide
- 5.0 ml of 5% aqueous potassium dichromate

Immerse tiny fragments (1 mm or smaller) for 6–18 h. Washing and disposal of used fixative are exactly as for Flemming's strong fluid. This mixture can also be used for post-fixation of small pieces that have been cut from larger glutaraldehyde-fixed specimens, to enhance the stability and improve the staining properties of membranous organelles such as mitochondria.

7. Fixation of thin layers of cells and cryostat sections

Thin layers of cells include tissue cultures, blood smears, cells that have been deposited on slides by centrifugation or by sedimentation of a suspension, and impression smears made by touching the freshly cut surface of an organ or tumour with a slide [23,40]. If cells are allowed to air-dry before fixation they may flatten considerably and appear unnaturally large. This change is least marked with blood films, in which the cells are supported by the drying plasma. Cells cultured on surfaces are often greatly flattened even while they are alive. It is generally desirable to immerse all preparations in fixative before they have dried by evaporation. The methods for fixing layers of cells are also applicable to mounted cryostat section of previously unfixed specimens.

7.1. Fixatives and procedures

As for a solid specimen, the correct fixative for a layer of cells is determined by the nature of the investigation. For example, minimal fixation in a neutral formaldehyde solution may be desirable for subsequent enzyme activity histochemistry, whereas a coagulant is likely to be more appropriate if dyes or immunohistochemical methods are to be used. A non-aqueous coagulant, with or without acetic acid, is commonly chosen.

The simplest way to apply a coagulant fixative is to immerse the slide or coverslip that bears the cells in methanol or ethanol, or to pour one of these liquids into a Petri dish containing cultured cells. Penetration and coagulation are completed in approx. 30 s.

Do not use solvents that are more hyhdrophobic than ethanol, e.g. acetone, with cultured cells because the solvent may dissolve the plastic substrate on which the cells were grown.

After fixation, store in 70% ethanol or methanol. Rinse in water before staining.

Unfixed cryostat sections, dried onto slides or coverslips, are often fixed by applying a drop of acetone and waiting until it evaporates. Many enzymic activities are preserved in such sections, but some are not [41].

Any of the neutral formaldehyde mixtures described earlier in this chapter may be used for minimal fixation. The cells should be immersed for 1–5 min, at either 4 °C or room temperature. Pilot experiments are needed to determine the optimum time. Generally this will be the longest that is compatible with demonstration of the enzymic activity. Minimally fixed cells are washed in three changes of a saline solution (do not use water) to remove the fixative.

References

1. Mori, S., Sternberger, N.H., Herman, M.M. and Sternberger, L.A. (1992) Histochemistry **97**, 237–241

2. Goodwin, A.E. and Grizzle, R.M. (1994) J. Histochem. Cytochem. **42**, 109–114

3. Bald, W.B. (1983) J. Microsc. **131**, 11–23

4. Lemke, C., Schwerdtieger, M., Pohlmann, I., Sammler, G. and Linss, W. (1994) Biotech. Histochem. **69**, 38–44

5. Pearse, A.G.E. (1980) Histochemistry: Theoretical and Applied, 4th edn., vol. 1., Preparative and Optical Technology, Churchill–Livingstone, Edinburgh

6. Ashwood-Smith, M.J. (1971) in Dimethyl Sulfoxide (Jacob, S.W., Rosenbaum, E.E. and Wood, D.C., eds), pp. 147–187, Marcel Dekker, New York

7. Lepault, J., Bigot, D., Studer, D. and Erk, I. (1997) J. Microsc. **187**, 158–166

8. Rosene, D.L., Roy, N.J. and Davis, B.J. (1986) J. Histochem. Cytochem. **34**, 1301–1315

9. Robinow, C. and Kellenberger, E. (1994) Microbiol. Rev. **58**, 211–232

10. Kok, L.P. and Boon, M.E. (1992) Microwave Cookbook for Microscopists. Art and Science of Visualization, Coulomb Press Leyden, Leiden

11. Login, G.R. and Dvorak, A.M. (1993) Scanning **15**, 58–66

12. Bernhard, G.R. (1974) Stain Technol. **49**, 215–224

13. March, J. (1977) Advanced Organic Chemistry: Reactions, Mechanisms and Structure, McGraw–Hill, New York

14. Baker, J.R. (1958) Principles of Biological Microtechnique (reprinted 1970, with corrections), Methuen, London

15. Baker, J.R. and Luke, B.M. (1963) Q. J. Microsc. Sci. **104**, 101–106

16. Walker, J.F. (1964) Formaldehyde, Reinhold, New York

17. Richardson, K.C. (1960) J. Anat. **94**, 457–472

18. Lillie, R.D. and Fullmer, H.M. (1976) McGraw–Hill, New York

19. Hayat, M.A. (1981) Principles and Techniques of Electron Microscopy. Biological Applications, University Park Press, Baltimore, MD

20. Paljarvi, L., Garcia, J.H. and Kalimo, H. (1979) Histochem. J. **11**, 267–276

21. Monsan, P., Puzo, G. and Marzarguil, H. (1975) Biochimie **57**, 1281–1292

22. Sabatini, D.D., Bensch, K. and Barrnett, R.J. (1963) J. Cell Biol. **17**, 19–58

23. Kiernan, J.A. (1999) Histological and Histochemical Methods: Theory and Practice, Butterworth–Heinemann, Oxford

24. Casselman, W.G.B. (1955) Q. J. Microsc. Sci. **96**, 203–222

25. Kiernan, J.A. (1985) Histochem. J. **17**, 1131–1146

26. Baker, J.R. (1946) Q. J. Microsc. Sci. **87**, 441–470

27. Elftman, H. (1954) J. Histochem. Cytochem. **2**, 1–8

28. Cammermeyer, J. (1960) Exp. Neurol. **2**, 379–405

29. Kiernan, J.A., MacPherson, C.M., Price, A. and Sun, T. (1998) Biotech. Histochem. **73**, 244–254

30. Carson, F.L., Martin, J.H. and Lynn, J.A. (1973) Am. J. Clin. Pathol. **59**, 365–373

31. Karnowsky, M.J. (1965) J. Cell Biol. **27**, 137A–138A

32. James, J. and Tas, J. (1984) Histochemical Protein Staining Methods (Royal Microscopical Society Microscopy Handbooks 04), Oxford University Press, Oxford

33. Moore, K.L., Graham, M.A. and Barr, M.L. (1953) Surg. Gynecol. Obstet. **96**, 641–648

34. Gregory, R.E., Greenway, A.R. and Lord, K.A. (1980) Stain Technol. **55**, 143–160

35. Fish, P.A. (1895) Trans. Am. Microsc. Soc. **17**, 319–330

36. Presnell, J.K. and Schreibman, M.P. (1997) Humason's Animal Tissue Techniques, Johns Hopkins University Press, Baltimore, MD

37. Churukian, C.J. (1997) Manual of the Special Stains Laboratory, University of Rochester Medical Center, Rochester, NY

38. Clark, G. (1973) Staining Procedures used by the Biological Stain Commission, Williams & Wilkins, Baltimore, MD

39. Baker, J.R. (1966) Cytological Technique, Methuen, London

40. Boon, M.E. and Drijver, J.S. (1986) Routine Cytological Staining Techniques. Theoretical Background and Practice, Elsevier, New York

41. Lojda, Z., Gossrau, R. and Schiebler, T.H. (1979) Enzyme Histochemistry. A Laboratory Manual. Springer-Verlag, Berlin

Tissue processing and sectioning

R.T. Allison

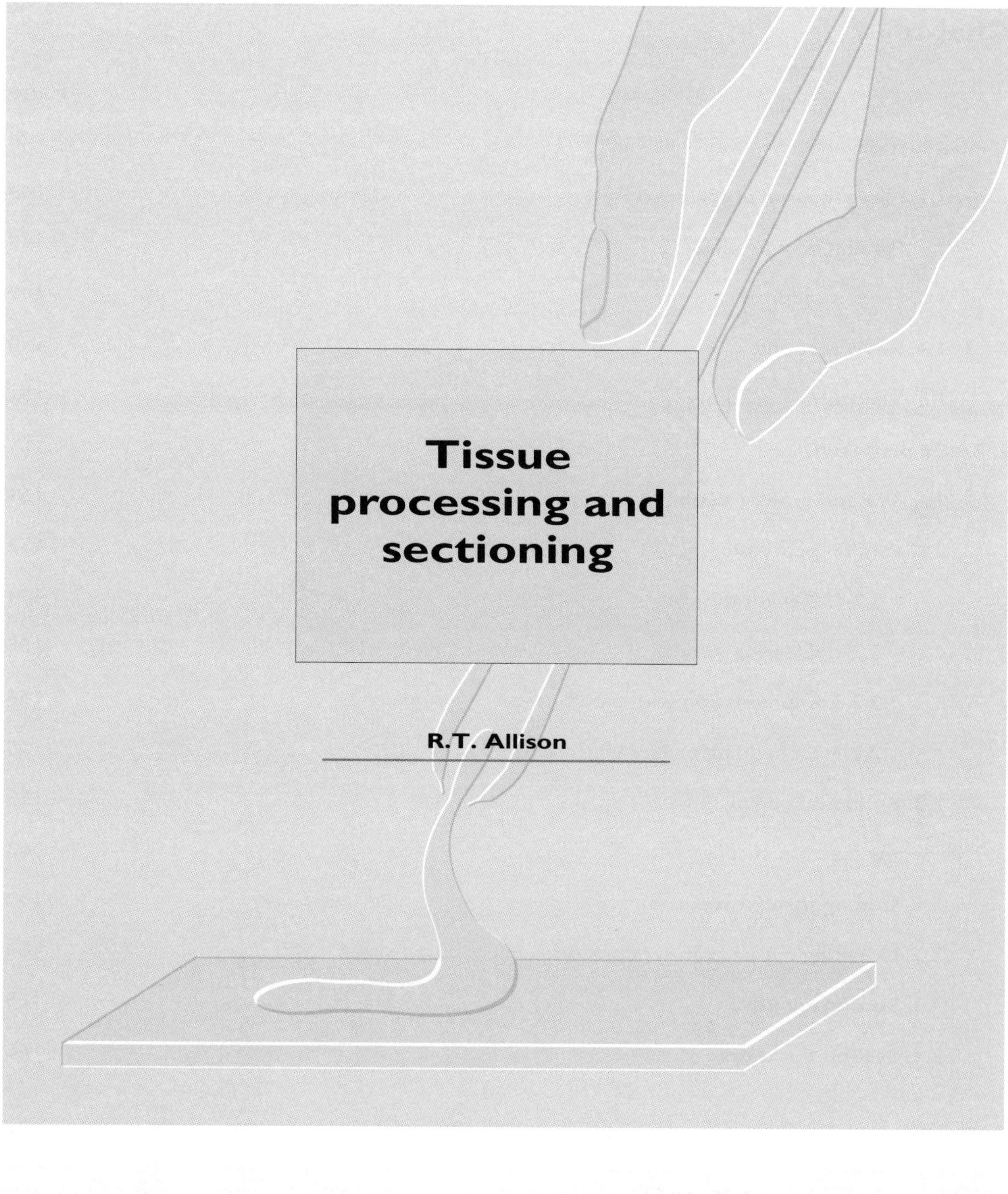

Chapter 7

1. Introduction

1.1. Microscopical preparations

Tissue processing has a single aim — the seemingly simple objective of facilitating the microscopic study of cells and tissues. For solid tissue (and the constituent cells), this is achieved by processing fixed tissue and cutting thin sections of it. Most commonly, these sections will be approximately a single cell thick. When required, dissociated cells may be studied by simply smearing them on microscope slides. Tissues may be viewed whole if thin enough for light to transmit (bladder, cultured monolayers). Thin sections may be prepared fresh or frozen without prior processing to study particularly labile or soluble substances. Unfortunately, not all cell and tissue components react optimally using a common preparatory method. Within certain broad parameters, a processing method which most closely favours the entity to be studied is chosen. For example, nuclear detail and tissue structure are best studied in 'paraffin sections', enzymes in 'frozen sections', basement membranes in thin sections and nerve axons in thicker ones.

By far the commonest method of tissue preparation is the procedure leading to a block of tissue embedded in paraffin wax. The objective of this is to provide the tissue with sufficient support to enable thin (3–20 μm) sections to be cut. Other methods with the same objective include cutting fresh unfixed (or fixed) tissue with a vibrating blade, freezing the tissue, or embedding it in a water-miscible wax or in a resin.

Paraffin sections are favoured because of ease of preparation, widest application and familiarity (the method has been around for nearly 200 years!). This means there is a huge body of knowledge based on the interpretation of sections prepared in this manner. Unfortunately, with roots in such a distant past, it is not the optimum method for some of the more analytical methods available to the modern microscopist.

1.2. Paraffin wax processing

Wax and water are not miscible; therefore tissue has to be dehydrated before it can be embedded in wax. Wax is, for practical purposes, immiscible with most dehydrants that have been tried (acetone, ethanol, methanol, etc.). An intermediary is therefore required. These are referred to as 'clearing agents' in recognition of the property of chloroform — an early choice — to render tissue somewhat transparent. They may also be referred to as 'antemedia' as they precede the embedding material. Finally, the tissue is infiltrated with molten paraffin wax before being cast in a block of the same material (embedding).

Causton [1] discusses infiltration with an example that serves equally well for wax processing. He defines the rate of diffusion as a function of the concentration of the infiltrating substance, the concentration already present at the point of interest and the diffusion coefficient, which will vary with different chemical groups within the tissue. The diffusion coefficient increases exponentially with increases in temperature, and it is also related to the molecular size of the diffusing substance.

The contraindications to paraffin wax processing relate particularly to the solvent properties of alcohol and of clearing agents, especially xylene. Clearing agents dissolve fats and simple lipids (compound lipids, e.g. myelin will resist these solvents).

Although fixation undoubtedly plays the major role in preserving and localizing cell and tissue entities, there are many antigens, for example, that remain demonstrable after identical fixation, but that are lost in different processing methods.

1.3. Frozen sections

Freezing can provide the rigidity necessary for microtomy but, unless performed carefully, causes considerable morphological disruption due to the formation, during freezing, of ice crystals within the tissue. Technically, frozen sections are a greater challenge than paraffin. A significant advance, originally devised to facilitate enzyme histochemistry, has been the use of a microtome in a deep freeze — the cryostat. The morphology of frozen sections, particularly if unfixed, is not always of the standard familiar in paraffin sections, but the tissue suffers less insults during preparation, and more chemical moieties are preserved, including virtually all antigens, enzymes and lipids. There is, however, always a need to be cautious with regard to localization in frozen sections that have not been 'post-fixed'.

1.4. Resin sections

Paraffin wax is insufficiently rigid for the preparation of the 'ultra-thin' sections used for electron microscopy (EM). Initially epoxy resins, and then methacrylates and polyesters under a variety of proprietary names, were used as embedding media for EM, with processing schedules similar in principle to those used with paraffin wax. The ease of use and good infiltration properties of methacrylate led to its adoption for preparing thin sections for light microscopy, particularly in the fields of renal and lymphoid diseases. Methacrylate sections are currently less popular for light microscopy, largely because of the advent of disposable microtome blades with which thinner paraffin sections can be cut.

There are particular problems associated with staining of methacrylate sections which are related in part to the impervious nature of the resin towards aqueous stains. Thin sections (typically 2–3 μm or less) contain only a small amount of substrate to stain. Perhaps the greatest benefit of the method has been the critically improved resolution of morphological features such as basement membrane and cell nuclei. Resin sections call for special microtomes, which may be used also, though less easily, for cutting paraffin sections.

1.5. Vibratome® sections

The Vibratome® (www.vibratome.com/sections) or vibrating microtome was devised to allow fresh or fixed tissue to be cut without freezing or processing, in the belief that it would produce 'life-like' sections. The principal benefits are those brought by the absence of both freezing artefacts and exposure to reagents other than the buffered saline or water in which sectioning takes place. Although sections less than 20 μm thick are achievable using the instrument, it is usually used to cut slices 50–500 μm thick, which may be cultured, investigated biochemically, fixed, stained or prepared for immunohistochemistry, with subsequent processing for light or electron microscopy. The Vibratome® is used in a wide range of settings and finds particular application in brain studies.

2. Tissue processing

Processing is the term used in histological laboratories to indicate the procedure for transferring a specimen from a fixative (typically an aqueous solution) to a wax or resin that is not miscible with water. Usually the tissue must be dehydrated, cleared, infiltrated with embedding medium and cast into a solid block.

2.1. Size and type of tissue

The *cri de coeur* of the cut-up area is "Too thick!". Despite such entreaties, tissue slices that are too thick are the commonest cause of processing failures. Those who cut up the tissue should be made to cut the sections. Tissue processing consists of diffusion and exchange of solvents in a finite time. The thicker the tissue slice, the longer it takes to process. Using automatic tissue processors, coping with individual variation of tissues is not an option, but thickness is only one dimension of a specimen. There is no limit on the other two, except the practicality of space within the cassette.

The dynamics of replacement diffusion are embodied in the rule that 'the thicker or denser the tissue, the longer it takes to process'. Dense tissues include decalcified bone, skin, central nervous tissue and fatty tissue, which may take twice as long as, for example, kidney tissue to process fully. The use of heated reagents in the processing cycle may speed up the process, but this may result in the hardening of muscle and tissues that contain much blood and/or collagen, thereby making them more brittle and difficult to section. Generally, a compromise must be sought for all the tissues being processed simultaneously.

2.2. Paraffin processing

For particularly dense (e.g. brain), tough (e.g. decalcified compact bone) or delicate (e.g. the tympanic membrane) tissues, it may be advisable to process tissue 'by hand'. However, most laboratories use 'closed' automatic tissue processors. These bring the advantages of convenience, versatility and enhanced safety to the procedure. Additionally, the processors will work overnight. It may take 2 or 3 days to process anything other than the smallest specimens by hand, because only 8–9 h are available in the working day.

A modern processing machine leaves the tissue in a retort into which the processing fluids, including molten paraffin wax, are pumped and withdrawn in sequence. The fluids may be agitated by rotation or tidal flow to enhance fluid exchange. Heat may be used at any stage to accelerate fluid exchange and reduce the processing time.

These sophisticated machines allow the operator to choose the time spent by the specimens in each of the processing stages. Therefore it is possible to process tissue urgently (in approx. 1 h) or over a more leisurely period

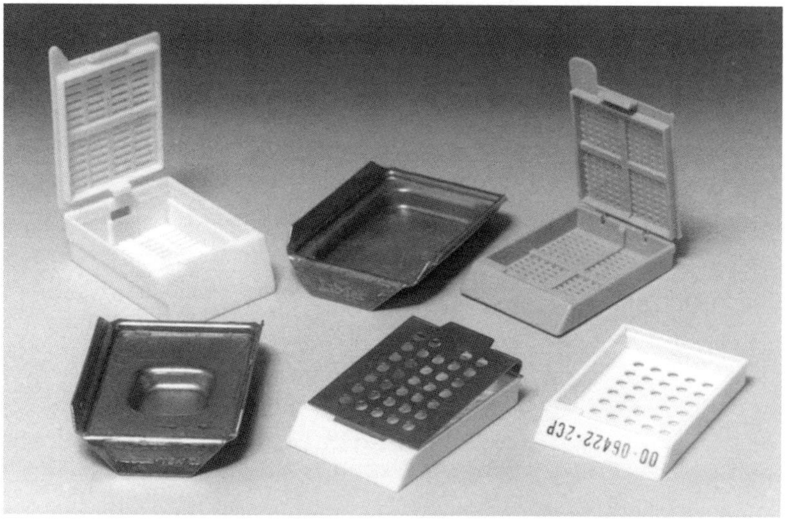

Fig. 1. Processing cassettes with reusable and integral lids and examples of embedding base moulds

of days. In routine diagnostic pathology, it is common practice to have a rapid processing program for urgent and/or small biopsies and an overnight program for 'routine' tissue. The following factors will influence the time taken for optimum processing:

(i) the thickness of the tissue sample;

(ii) the clearing agent used;

(iii) the use of vacuum and agitation;

(iv) the nature of the tissue.

The absolute necessity for tissue identity is maintained by the use of cassettes with detachable lids (Fig. 1). When a specimen is placed in a cassette, its identifying number is inscribed on the outside using a graphite pencil or an automatic cassette numbering machine (which burnishes the number, etc. into the plastic). Ordinary inks should not be used because they are soluble in organic solvents. These cassettes have the added advantage of limiting the thickness of the tissue block, a feature critical to satisfactory processing.

Automatic processors are flushed to clean the retort and transport chamber of residual wax using a cycle comprising a warm clearing agent (xylene) followed by ethanol. Many laboratories find it useful to follow this with a warm water rinse. The maximum number of tissue-containing cassettes that may be processed in a single run varies according to the manufacturer, but it may be measured in hundreds. Periodically, it is necessary to replace the processing fluids as they become exhausted. The frequency of this will depend on the amount of tissue processed; if used daily it is recommended that they are changed weekly. A recommended procedure is to discard the 'oldest' fluid where practical and to move subsequent fluids 'up' the processing cycle.

2.2.1. Dehydration

The commonest dehydrants are ethyl alcohol (ethanol) and cheaper substitutes such as isopropyl alcohol (isopropanol; propan-2-ol) or absolute industrial methylated spirit (74 OP). Methylated spirit is a popular dehydrant in the U.K., where it is not subjected to the same customs and excise control as ethyl alcohol. There are conflicting statements on the effects of dehydrants on subsequent eosin staining. It is commonly believed, and it is possibly a truism, that plunging a specimen directly into absolute alcohol from an aqueous fluid (such as a fixative) will cause structural damage to tissue. It is recommended that graded dilutions of alcohol be routinely used: successively 50%, 70%, 95% and 100% (v/v). Tissue is tolerant of dilute alcohol and may be stored in it, particularly after fixation in intolerant fixatives. Alcohol is itself a fixative and has a probable role in the further fixation of tissue during processing. In contrast with formaldehyde, which cross-links protein groups, alcohol is a coagulant (precipitant). There is evidence that this action brings additional benefits in exposing further reactive tissue sites within the tissue. Alcohol as a primary fixative enhances the demonstration of many tissue antigens by immunocytochemistry.

Dehydration is achieved by simple exchange diffusion, which must be completed by the time tissue is transferred to a clearing agent. The time will depend upon the volume ratios, the thickness and density of the tissue blocks, agitation and, some would say, temperature. For routine purposes, tissue blocks no more than 4–5 mm thick should be completely dehydrated following treatment with 50%, 70% and 90% alcohol and three changes of absolute alcohol, each lasting for a least 1 h. These are the solutions that can be 'moved up' during weekly refreshing of the liquids in the processor. There are those who recommend adding anhydrous copper sulphate (a white powder that turns blue in the presence of water) to absolute alcohol, both to aid dehydration and to indicate exhaustion of the alcohol, though its use is of questionable value.

Specimens that have been fixed in alcohol-containing fixatives, such as Carnoy's fluid, can be moved immediately into a higher grade alcohol (such as 90% alcohol).

A novel system of chemical dehydration using 2′2-dimethoxypropane has been investigated by Conway and Kiernan [2] and it was found to give satisfactory results, including unimpaired RNA staining. Whether it affects the performance of closed tissue processors is undetermined, but it does reduce the number of fluid changes required and less 2′2-dimethoxypropane is needed, compared with alcohol in traditional processing, thereby making it a cost-effective alternative.

2.2.2. Clearing

Aromatic hydrocarbons, especially xylene, are the most commonly used clearing agents. Toluene has lost some of its popularity because it poses a greater health risk. Many laboratories find it economical (and environmentally friendly) to have xylene recycled by distillation, either in-house or commercially. Historically, benzene was favoured for routine use, and both chloroform and cedarwood oil were preferred for particularly delicate tissues.

Chloroform may have carcinogenic and/or teratogenic potential, and despite undoubted advantageous qualities, it is used less frequently than it was in the past. Cedarwood oil is therefore regaining some popularity as the most 'gentle' clearing agent for delicate tissues. Both of these clearing agents are especially useful in manual processing. Schedules include at least one overnight stage and either of these clearing agents is preferable to xylene, which may cause excessive hardening of the tissue after prolonged exposure. As may be expected, cedarwood oil, a viscous liquid, is more difficult to replace with molten wax in tissue. It is recommended that this substance not be used in automatic tissue processors.

More recently, 'environmentally friendly' clearing agents have been introduced, but not all have maintained their popularity. These were derivatives of natural food oils and they were thought to have no toxicity. Some of these liquids, however, have more offensive smells than xylene, and the assumption that they could replace xylene for all uses proved erroneous. This is true particularly in the case of 'clearing' stained slides prior to mounting (see Chapter 8).

2.2.3. Impregnation with paraffin wax

All paraffin waxes are mixtures of straight-chain hydrocarbons produced by the distillation of crude oil. Different sources of oil affect the properties and quality of the waxes obtained. Although categorized by their melting point, in chemico-physical terms, the plastic point is the more important [3]. In general, it may be assumed that the higher the given melting point the harder the wax. For practical purposes, waxes used in histological techniques have melting points in the range 55–60 °C. There have been many and varied attempts to improve the infiltrating and sectioning properties of paraffin wax. Successful section cutting is based on personal preferences that develop with experience and these may extend to personal preferences in the choice of the paraffin wax.

From experience, it can be said that a melting point of 58 °C is generally acceptable, that few additives have won universal approval, that wax crystal size is of marginal significance, that there are benefits when cutting certain tissues in using different wax formulations for infiltrating and embedding (see below), that environmental temperatures will affect the cutting characteristics of the wax, and that 'no magic' wax exists to rectify the effects of poor processing.

Paraffin wax should be clean, it should be devoid of foreign material that could damage the blade, and it should have good ribboning properties. Advantages may be gained by the use of wax containing dimethyl sulphoxide for infiltration purposes and one containing plastic polymers for final embedding of tough tissues or if trying to cut thin (2 μm) sections.

Suitable schedules for overnight processing, for rapid processing and small biopsies and for 'hard' processing are given in Tables 1, 2 and 3 respectively.

Table 1. Overnight paraffin processing

Reagent	Time (h)	Temperature
Alcohol (50%)	1	Room temperature plus agitation
Alcohol (70%)	1	Room temperature plus agitation
Alcohol (90%)	1	Room temperature plus agitation
Absolute alcohol (100%)	1	Room temperature plus agitation
Absolute alcohol (100%)	1	Room temperature plus agitation
Absolute alcohol (100%)	1	Room temperature plus agitation
Xylene	1	Room temperature plus agitation
Xylene	1	Room temperature plus agitation
Xylene	1	45 °C plus agitation
Wax	1	60 °C plus vacuum
Wax	1	60 °C plus vacuum
Wax	1.5	60 °C plus vacuum

2.2.4. Embedding or blocking out

A fully wax-impregnated specimen is cast into a block using a suitable mould, either disposable or re-usable. The Tissue-Tek™ cassette system is easy, convenient and in widespread use. The metal mould is filled with molten wax and, before it solidifies, the wax-infiltrated object is placed in the mould with the surface to be cut facing down. It is helpful to have the metal mould on a cool surface until the setting wax holds the specimen in the desired orientation. The numbered cassette is placed on top, and both mould and cassette are topped up with molten wax. Various mould sizes are available. It is possible to improvise by making moulds from grease-proof paper or thin card.

Ideally the mould is then placed on a chilled surface to solidify. Studies on the crystalline nature of paraffin wax have shown remarkably little variation between products [4]. However, many believe that a microcrystalline structure confers advantages during cutting, and rapid cooling from one surface certainly enhances the formation of small crystals [5]. Comments on the penetration of smaller crystals, occasionally seen in the literature, are nonsense (molten paraffin is a liquid).

Table 2. Rapid paraffin processing for small tissue blocks

Reagent	Time (min)	Temperature
Alcohol (50%)	15	Room temperature plus agitation
Alcohol (70%)	15	Room temperature plus agitation
Alcohol (90%)	15	Room temperature plus agitation
Absolute alcohol (100%)	20	Room temperature plus agitation
Absolute alcohol (100%)	20	Room temperature plus agitation
Absolute alcohol (100%)	30	Room temperature plus agitation
Xylene	15	Room temperature plus agitation
Xylene	20	Room temperature plus agitation
Xylene	30	45 °C plus agitation
Wax	30	60 °C plus vacuum
Wax	30	60 °C plus vacuum
Wax	30	60 °C plus vacuum

Table 3. Hand processing

Reagent	Time (h)*	Temperature
Alcohol (50%)	1 (10.00–11.00)	Room temperature; agitate frequently by hand
Alcohol (70%)	1 (11.00–12.00)	Room temperature; agitate frequently by hand
Alcohol (90%)	1 (12.00–13.00)	Room temperature; agitate frequently by hand
Absolute alcohol (100%)	1 (13.00–14.00)	Room temperature; agitate frequently by hand
Absolute alcohol (100%)	1 (14.00–15.00)	Room temperature; agitate frequently by hand
Absolute alcohol (100%)	1 (15.00–16.00)	Room temperature; agitate frequently by hand
Chloroform	1 (16.00–17.00)	Room temperature; agitate frequently by hand
Chloroform	Overnight	Room temperature
Chloroform	1	Room temperature plus agitation
Wax	1	60 °C plus vacuum
Wax	1	60 °C plus vacuum
Wax	1	60 °C plus vacuum

Extend each wax stage by 30 min if vacuum not available.
* Values in parentheses are show a suggested time-scale for such a procedure.

2.3. Resin embedding

Traditionally, acrylic (methacrylate) embedding has been used for light microscopy of tissue, while epoxy (e.g. araldite, epon) is recommended for electron microscopy The introduction of 2-hydroxyethyl methacrylate (HEMA) usurped the popularity of glycol methacrylate and polymethyl methacrylate. One reason is that HEMA is tolerant of water, which leads to stronger staining of the sections by conventional methods. It is rarely necessary to inhibit spontaneous polymerization of HEMA, although each batch should be tested before use. Glycol methacrylate contains hydroquinone as an inhibitor of polymerization, and this has to be removed before use.

With carefully controlled polymerization, it is possible to use the acrylic variants L R White® and, more particularly, L R Gold® [6] for both light and electron microscopy, and this can be advantageous in immunocytochemistry.

The two phases of resin equivalent to molten and solidified wax are unpolymerized (liquid) and polymerized (solid). Following complete infiltration with unpolymerized resin, the tissue is infiltrated with resin containing a chemical polymerization agent. In general, smaller pieces of tissue will be processed and cut in resin rather than in paraffin wax because unpolymerized resin infiltrates less effectively than wax and because large resin sections are more difficult to cut and handle.

Caution *Unpolymerized resin can easily induce skin sensitivity in some individuals and benzoyl peroxide it a significant oxidizing agent.*

Method 1. Embedding in HEMA

Materials

Solution 1

- 80 ml of HEMA
- 8 ml of butoxyethanol
- 1 g of benzoyl peroxide

Solution 2 (activator)

- Polyethylene glycol 400 (15 parts)
- *N,N*-Dimethylaniline (1 part)

Procedure

1. Dehydrate fixed tissue completely in graded alcohols, as in Table 2.
2. Place in Solution 1 for 2 h.
3. Replace with fresh Solution 1 overnight.
4. Embed in mixture of Solution 1 and Solution 2 (42:1 by volume).
5. The mixture containing the tissue should be placed in a mould together with the stub with which it will be attached to the microtome (Fig. 2).

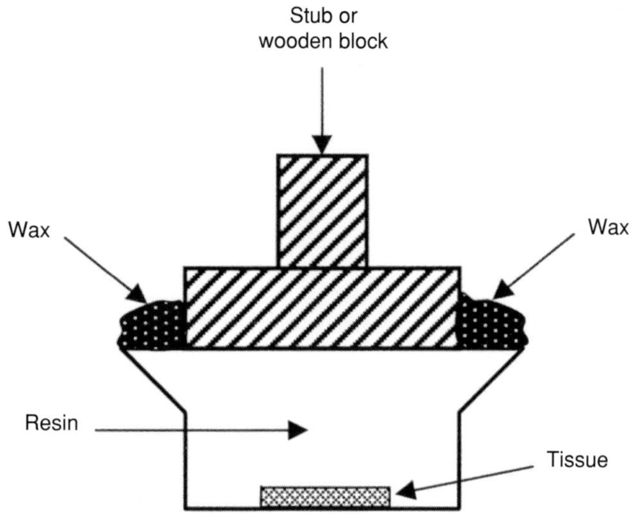

Fig. 2. 'Anaerobic' polymerization of resin

Alternatively, resin blocks may be stuck onto hardwood blocks of appropriate size.

6. Optimize polymerization under anaerobic conditions by sealing the exposed resin with molten paraffin wax. Commercial kits are available for this.

7. At room temperature, polymerization should be complete within 4–6 h.

   Polymerization is an exothermic reaction. Butoxyethanol and polyethylene glycol are included as plasticizers, which improve the cutting qualities of the resin.

3. Microtomy (section cutting)

The history of microtomy is littered with novel cutting machines (microtomes), but the base sledge (for paraffin) and rotary (for both paraffin and resin) have survived as the most popular choices for 'processed' tissue. Specialist microtomes, most commonly based on the rotary, are available for cutting thin resin sections, and there are various hybrids such as the freezing/cryostat microtome (see below) and the sliding microtome. There is no substitute for experience for success in microtomy; it is an individual skill. Variables in the quest for success (not including those associated with tissues) include, first and foremost, the sharpness of the blade, followed by the angle of the blade relative to the block, the bevel of the cutting edge, the speed of cutting, atmospheric conditions (humidity, temperature, static electricity, etc.), the temperature of the block and the section thickness. Control of these variables will be acquired with experience.

3.1. Cutting paraffin sections

The following tips may prove helpful when cutting paraffin sections

- Use disposable blades.
- Ensure that the microtome is stable and that the blade/knife and block are held firmly.
- The knife angle (tilt) should generally be greater for hard tissue than for soft.
- Vary cutting speed according to the tissue (determining the appropriate cutting speed can be gained with experience). In general, it can be assumed that softer tissues are more easily sectioned at low speeds and harder ones at higher speeds.

> ***Note*** *Cutting too rapidly will compress sections. It is difficult to maintain the necessary rhythm when cutting too slowly.*

- Dampen the block surface before cutting (some workers use a paper tissue etc. for this, but this author prefers to leave the block face-down on melting ice prior to cutting). It should be noted that some tissues swell in water, even after embedding in wax. Material swelling from the surface of the block can be 'trimmed off'.
- Where static electricity is a problem, consider the clothing being worn — some synthetic materials are notorious for inducing static electricity. Where atmospheric problems are the cause, they may be alleviated by placing a Bunsen burner, with the pilot lit, safely in the vicinity of the knife.
- Before sections can be cut, the block must be trimmed off until the covering wax is removed and a complete surface of the tissue is exposed. This is done coarsely with the microtome by cutting relatively thick sections from the surface. Most microtomes have features to help facilitate this, but it may be done by advancing manually between each cut or by setting the section thickness at 15–20 μm. Cutting sections that are too thick at this stage can damage the tissue by tearing pieces from the surface. Once trimmed, the blocks are placed face down on the ice tray.

Whether using a rotary or a sledge microtome, the following guidance is relevant, but the goal is to obtain a 'ribbon' of sections.

- Clamp the knife or blade-holder firmly in the microtome, ensuring the desired angle relative to the block surface.
- Place the tissue block securely in its holder and manually bring the surface of the block into near contact with the blade edge. The surface of the block and the leading edge of the block should be parallel to the blade edge. The former will obviate the need for excessive trimming, while the latter helps ensure a straight ribbon.
- Set the microtome to cut sections of the desired thickness, usually 5 μm (see below).
- Cut the first section, which should slide over the knife surface. Hold this section with a dampened fine brush or fine forceps and repeat the motion until a ribbon of sufficient length is achieved, with each section sliding across the cutting edge in turn and sticking to the previous section. Occasionally, this process may be helped by gentle, open-mouthed breathing on the block surface during the cutting action.

- Carefully release the last section from the blade edge by a gentle downward stroke with a second brush, taking care not to brush across the easily damaged edge of the blade.

3.2. Mounting cut paraffin sections onto microscope slides

The ribbon is floated onto a warm waterbath, set approx. 10 °C below the melting point of the chosen wax. The free end of the ribbon is allowed to touch the water and the ribbon is allowed to gently settle on the surface using a simultaneous dragging action. This enables the sections to expand to their full breadth, thereby recovering from the slight compression that occurs during cutting. Individual sections are separated by touching either side of the junction between adjacent sections simultaneously with a slightly closed pair of fine forceps and then allowing them to come apart gently. Clean forceps help to ensure that they do not stick to the wax (practice is required to master this technique). Alternatively, the ribbon of sections may be laid on a piece of black card and sections separated with a scalpel before floating them out individually (Fig. 3).

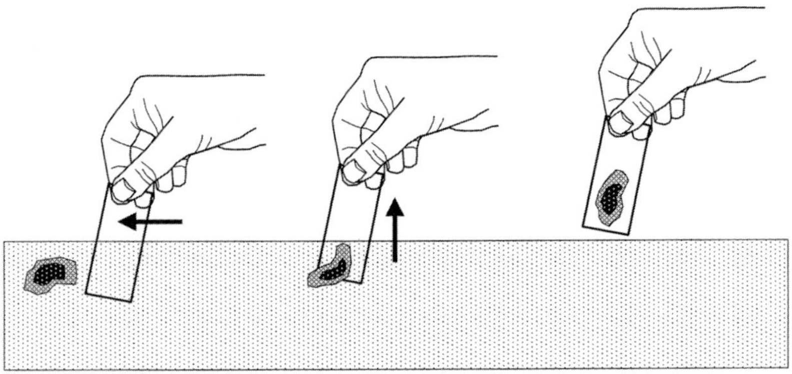

Fig. 3. Mounting sections onto slides from the floating-out bath

A slide should be introduced into the water so that it approaches the floating section from the side. When it touches the section it should be withdrawn gently from the water bath. If necessary, the section can be orientated on the slide while still wet, using the brush or forceps. This is often impossible if using charged slides (see below).

The section must be dried on the slide; the drying method depends on subsequent staining procedures. For many purposes this is achieved by warming the slide to a temperature just below the wax melting point for times varying from 30 min to overnight, depending on the planned staining. For some purposes it is considered beneficial to dry at room temperature, for example, when seeking to demonstrate labile antigens immunohistochemically. Some automatic staining machines include a drying stage. Hotplates are another alternative. Excess water is drained immediately from the slide before it is placed face down over the hot surface for 15 min.

3.3. Section adhesives

Not all tissue sections adhere equally well to slides (decalcified tissue may detach easily), and some staining procedures are more aggressive than others (methods using ammoniacal silver solutions and immunocytochemical procedures, for example). Schemes devised to overcome these problems can be divided, essentially, into two groups. The traditional approach is to provide an additional protein layer, such as albumen or gelatin, between glass and tissue. More recently, substances have been used that impart a positive electric charge to the glass surface. 3-Aminopropyltriethoxysilane is commonly used, as are proprietary products of uncertain identity.

3.4. Section thickness

For routine morphological studies, sections are cut at 4–5 μm, which is usually also suitable for *in situ* hybridization. However, for PCR investigations on archived paraffin wax-embedded tissue for example, a thickness of 10 μm or even 20 μm obviously gives more substrate to work with. It is important to be aware that nominal section thickness, as indicated by the microtome setting, is unreliable. Serial sections may vary in thickness, as may areas of the same section. Importantly, the thicknesses of nuclei may vary, not only because of fluctuations in the advancing mechanism of the microtome, but also because, as Helander [7] has clearly shown, the cutting edge of a blade may slice through a nucleus or be deflected over or under it during cutting. This may be a serious consideration in quantitative studies.

3.5. Frozen sections

Two methods, the use of the traditional freezing microtome and the cryostat, are available. The traditional freezing microtome is used primarily for cutting free-floating sections for the demonstration of lipids (which would be lost during processing through solvents) and for some stains applied to nervous tissue. Cryostat sectioning, devised originally for enzyme histochemistry on fresh unfixed tissue, now has more catholic appeal.

Frozen sections are more easily cut from unfixed than from fixed tissue. Freezing must be achieved as rapidly as possible to avoid artefacts, primarily ice crystals, which form during the freezing process and distort tissue architecture. The best method is to plunge small pieces of fresh tissue into isopentane that has been cooled with liquid nitrogen to approximately −160 °C for 1–3 min. Place the specimen on a small piece of foil before putting it into the isopentane. This helps freezing and also preserves the orientation of the object.

Frozen sections are frequently cut at 8–10 μm, but may be considerably thicker (20–80 μm) for nervous tissue. Unmounted sections are floated in the staining solutions (they transferred between each solution with a fine brush or glass 'hockey stick'). Finally, they are floated onto a glass slide, the excess water is drained off and the preparation is air-dried. It may be mounted in an aqueous medium (lipid stains, immunofluorescence) or dehydrated, cleared with xylene and covered using a resinous mountant.

The cryostat is essentially a tool to allow cutting of sections in a 'deep-freeze'. The temperature of the cabinet and of the knife is controllable and is normally in the range −10 °C to −30 °C. Specimens may be frozen as described above, or by mounting on a metal holder (with water or a sugar solution as adhesive) and then placing on a 'heat-sink' block, which is within the cryostat but maintained at a much lower temperature than the microtome and the atmosphere of the cabinet. Individual sections (not ribbons) are cut in much the same way as are paraffin sections, but with the important difference that they will not slide flat over the blade. They require the use of an 'anti-roll plate' which is in close apposition to the blade surface and between which the section is naturally guided as it is cut. Otherwise the section would simply curl into a roll at the knife edge.

When a section has been cut, the anti-roll plate is lifted and a slide or coverslip is brought into contact with the section. The section immediately binds to the slide and is ready for staining. Post-fixation of the section is possible if required, for example to maintain localization.

3.6. Vibratome® sections

Like all other microtomes, the Vibratome® may be called upon for sectioning a wide variety of tissues. The machine uses standard razor blades which are made to vibrate at 50–60 Hz as the specimen passes over the blade, providing a saw-like cutting action. The amplitude of vibration and the speed of advance are variable. The best results are usually obtained with slow advance and large amplitude. Controls are also provided for setting section thickness and, in some models, the temperature of the bath.

The specimen is stuck with cyanoacrylate glue to a block. This is held by a clamp in a bath of liquid, which may be a physiological saline solution (for unfixed tissue) or water (for fixed tissue). The sections float into the bath and are collected with a brush or a glass pipette. They can be mounted onto slides or transferred to vessels appropriate for culturing, metabolic studies, immunohistochemistry or other purposes. It is seldom possible to obtain sections less than 50 μm thick, but this is advantageous in many investigations, especially of the nervous system.

4. Troubleshooting

Problem	Diagnosis
Scores in sections	Damaged blade edge; calcium or foreign body in block; dirt in wax.
Sections curl up on blade as cut	Blunt blade; tilt of blade too great.
Sections alternately thick and thin; each section has zones parallel to blade edge (chatters)	Microtome screws need adjusting and tightening; tilt of blade too great; tissue very hard (needs softening); blunt blade.
Sections crumble on cutting	Blunt blade; wax too soft (try applying ice to surface); wax contaminated by water or clearing agent (replace).
Ribbon of sections curved	Block edges not parallel to each other; block edge not parallel to blade edge.
Breadth of section less than block; creases will not come out	Compression of block (clean back of blade); wax too soft (try higher m.p. wax); blunt blade.

References

1. Causton, B.E. (1980) Proc. R. Soc. **15**, 185–189

2. Conway, C. and Kiernan, J.A. (1999) Biotech. Histochem. **74**, 20–26

3. Allison, R.T. (1979) Med. Lab. Sci. **36**, 359–372

4. Allison, R.T. (1978) Med. Lab. Sci. **35**, 355–363

5. Demster, W.T. (1943) Mich. Acad. Sci. Arts Lett. **29**, 251–264

6. Newman, G.R. and Hobot, J.A. (1987) J. Histochem. Cytochem. **35**, 971–981

7. Helander, K.G. (1983) J. Microsc. **132**, 223–227

Staining and counterstaining to show structural features

John A. Kiernan

Chapter 8

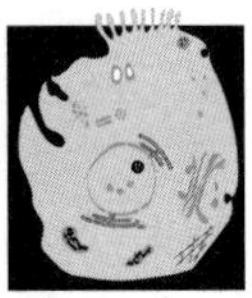

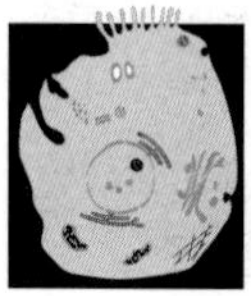

1. Introduction

This chapter contains instructions for some staining methods that make use of dyes to show the microscopic structure of cells and extracellular materials in sections. Dyes are coloured compounds that can bind to insoluble substrates. In sections of organs or thin layers of cells, the substrates include large molecules such as nucleic acids, proteins, carbohydrates, lipids (mainly in frozen sections) and other organic macromolecules such as amyloid and lignin. Some inorganic ions, notably calcium and iron, can also be stained with dyes *in situ*.

There are many modes of attraction between dyes and their substrates, and these determine the specificity of staining of different structures. For detailed discussion of the principles of staining, consult the works of Baker [1], Horobin [2,3], Kiernan [4] and Lyon [5]. The simplified explanations that accompany the following methods are based on attempts to reconcile widely divergent theories of staining.

The methods have been selected for their technical simplicity, but some thought is required when carrying out even the simplest staining procedure. It is frequently necessary, for example, to vary the timing of the stages in a method. The pH of a reagent may sometimes need to be changed, and so (although rarely) may the concentration of a dye. Dye binding is greatly influenced by the chemical changes that follow fixation and, in some cases, by the exposure of specimens or sections to solvents.

The aim of the first three groups of techniques is to display structural and simple chemical features of tissues. The fourth group involves the use of counterstains. These are used to impart colours that contrast with the results of other procedures such as immunohistochemical staining or *in situ* hybridization.

2. General instructions for staining

The most important rule of staining is to understand what you are doing. Read the short explanation of how each procedure works, read through the instructions (including these introductory paragraphs) and make sure you have all the required solvents and other chemicals to hand. If you still do not understand all the stages of the method, read its references or look it up in a textbook of histological technique [4,6–9].

Label all slides before staining. Use an ordinary graphite pencil to write on the ground glass at the end of the slide, or a diamond scriber to scratch plain glass. As well as identifying the preparation, the label shows which surface of the slide carries the sections, minimizing the chance of putting the coverslip on the back of the slide.

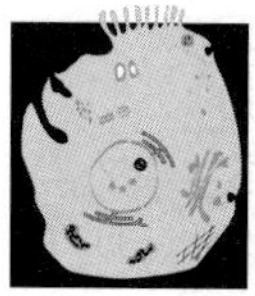

Small numbers of slides are moved with stainless steel or plastic forceps through solutions in small vessels such as Coplin jars or other dishes that hold 50–100 ml (Fig. 1). Larger numbers are carried in staining racks through 200–400 ml rectangular tanks with metal or glass lids. It is important to shake each slide or rack thoroughly and tap it on bibulous paper before moving it into the next vessel. This delays the contamination that will eventually occur with repeated use of the liquids.

Most dye solutions can be re-used repeatedly and can be stored for months or years, but they all fail eventually, so it is advisable to label the bottle with the date that the solution was made and to replace any old solution that seems to stain sections weakly or abnormally. Solutions of some dyes (e.g. eosin Y) serve as culture media for fungi; replace them if they contain visible mould or if fungal hyphae or bacteria are seen in stained sections. (They appear as thin filaments, lumps slightly smaller than cells, or fine granular dirt; they may be colourless, brown or strongly stained.) Alcohols and xylene can be re-used approximately 12 times. Change them if they show even slight turbidity or contain droplets of an immiscible liquid. A little colour does not matter in alcohols being

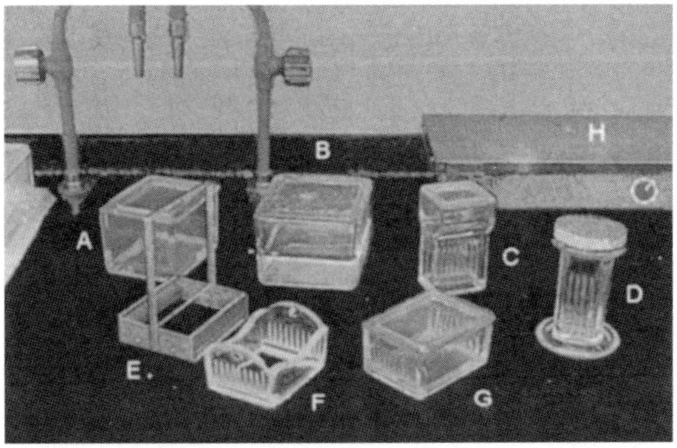

Fig. 1. Vessels and other items used for staining

(A) 200 ml tank with glass lid; (B) 400 ml tank with glass lid; (C) vertical staining jar with a glass lid and slots for eight slides; (D) a Coplin jar with plastic screw cap and slots for five slides; (E) stainless steel staining rack that holds 24 slides; (F) glass rack with slots for ten slides; (G) horizontal staining jar with slots for ten slides and glass lid; (H) electric hotplate, used for drying slides that have either newly cut sections or recently applied coverslips.

used for dehydrating stained slides, but the final container of absolute alcohol should be colourless.

In this account, alcohol means ethanol, but methanol or isopropanol (propan-2-ol) can usually also be used. Industrial methylated spirit or denatured ethyl alcohol is treated as 95% ethanol.

Water refers to distilled or deionized water unless otherwise indicated. Acidified water is distilled or tap water with approx. 5 ml of glacial acetic acid per litre, and slightly alkaline water is made by adding a pinch (approx. 50 mg) of calcium hydroxide or sodium acetate, or approx. 5 ml of saturated aqueous lithium carbonate (stock solution) to each litre. Clearing of dehydrated sections before addition of the coverslip is done by passing the slides through two changes of xylene (or any other non-polar solvent that is miscible with 100% alcohol and with resinous mounting media). Mounting media are usually proprietary products.

Cytoseal-60 (VWR Canlab) and Entellan (BDH), which are both based on methacrylate polymers, are recommended mounting media.

Fig. 2 is a flow chart showing the steps involved in staining a batch of slides, starting with either mounted sections in paraffin wax or mounted frozen (or cryostat) sections, and ending with a permanent coverslipped preparation.

Keep an old microscope in the laboratory, and always examine stained sections before coverslipping, to see if the colours are as they should be. This is particularly important when staining a large batch of slides because removal of coverslips and mounting medium (see Section 7) is a slow business and can damage the sections.

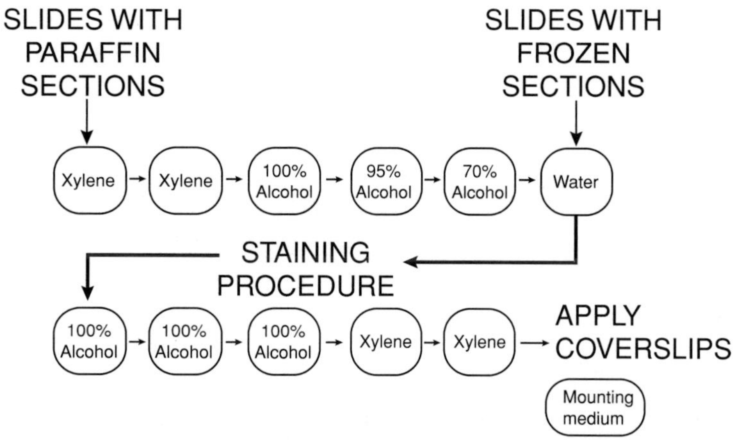

Fig. 2. Flow-chart for rehydrating paraffin sections, staining, dehydration, clearing and coverslipping with a resinous mounting medium

Frozen sections do not require rehydration before staining.

2.1. A note on buying dyes

When buying a dye for staining, check that its CI (colour index) number is correct. This identifies the right compound but does not guarantee its efficacy. Whenever possible use certified dyes, which are identified as such in catalogues and on the bottles. Certification is granted to batches of dyes submitted by vendors for testing by an independent non-profit organization, the Biological Stain Commission, in the U.S.A. All the dyes mentioned in this chapter are sold by Aldrich, Sigma and Fluka. The other chemicals needed are available from all laboratory supply houses.

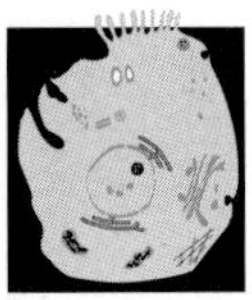

3. General oversight methods

The first two methods in this group stain cell nuclei in shades of blue or purple, and other tissue components in shades from orange–red to pink–violet. They are appropriate for 4–15 μm thick sections that are mounted on slides. The third method is for mounted semi-thin (0.5–1.0 mm) sections taken from small plastic-embedded specimens, such as those from which ultrathin sections are cut for electron microscopy.

3.1. Staining with eriochrome cyanine R

This simple technique [9a] provides two colours from one dye and a ferric salt at pH 1.5. This solution contains an anionic red $[Fe_2H\text{–}(dye)]^-$ complex that binds to most components of the tissue. In nuclei and myelin this loses a proton, perhaps to the strongly basic histones in nuclei and to the phospholipid bases in myelin, and becomes a blue complex, thought to be $[Fe_2\text{–}(dye)]^{2-}$ [9b] .

The result resembles staining by the much more common, but more difficult to use, alum–haematoxylin and eosin (H&E) method. The procedure given here differs in some details from the method originally described by [9a].

Method 1. Staining with eriochrome cyanine R

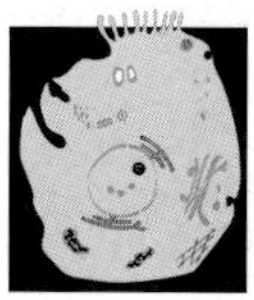

Materials

- Distilled or deionized water
- 95% Alcohol
- 100% Alcohol
- Xylene
- Resinous mounting medium
- Staining solution
  - 0.14 g of ferric chloride ($FeCl_3 \cdot 6H_2O$). (It is easier to make 100 ml of 0.21 M $FeCl_3$ and use 2.5 ml of this solution.)
  - 1.0 g of eriochrome cyanine R (CI 43820; Mordant blue 3). This dye is also called chromoxane cyanine R and solochrome cyanine R.
  - 2.5 ml of concentrated (96%) sulphuric acid.
  - Water (to make volume up to 500 ml).

This solution is stable for at least 10 years and can be used repeatedly. It requires occasional pH adjustment (the pH must be 1.5 in order to obtain the correct colours in the tissue). Add a few drops of either 1.0 M hydrochloric acid or sulphuric acid, or 1.0 M sodium hydroxide to bring the pH into the range 1.46–1.54. Adjustments are seldom necessary for solutions more than 2 weeks old, but are sometimes needed if a year has passed since the last adjustment.

Procedure

This method is for mounted sections 4–15 mm thick. It works after most fixatives.

1. Dewax the sections and bring them through graded alcohols into water (not needed for frozen sections).

2. Stain slides bearing hydrated sections for 3 min.
3. Wash in three changes of water (must be distilled or deionized) for 20 s each time, with agitation.
4. Dehydrate in three changes of 100% alcohol.
5. Clear in two changes of xylene and apply coverslips, using a resinous mounting medium.

Results

Nuclei stain blue–purple; cytoplasm stains pink to red; collagen stains mostly red, but some fibres appear purple; erythrocytes stain orange–red; myelin stains blue–purple. The balance of colours is critically dependent on the pH of the staining solution. If this is too acidic, red will predominate; if it is not acidic enough there will be too much purple and blue staining.

3.2. Alum–H&E

Despite the common name of this technique, haematoxylin is not a dye and does not stain anything. Nuclei are stained blue with haemalum, which is an acidic solution containing aluminium ions, haematein and a highly polar substance such as glycerol or ethylene glycol that helps to keep the dye in solution. Haematein, a yellow dye, is generated by the oxidation of haematoxylin, which is a colourless compound extracted from the heartwood of a tropical American tree, *Haematoxylon campechianum* or logwood. The oxidation of haematoxylin, known as 'ripening', occurs slowly in the presence of air and almost instantly with a stronger oxidizing agent such as sodium iodate. Eventually haematein is further oxidized to oxyhaematein, an orange–yellow compound that lacks useful staining properties. Over-oxidation occurs more rapidly with added oxidizing agents, than with atmospheric oxygen [10]. Haematein forms strongly coloured dye–metal complexes. Those with aluminium are red in acid solutions and blue in a neutral environment.

A solution containing a large excess of Al^{3+} ions over haematein molecules selectively stains nuclei. Although the blue dye–metal complex can bind to DNA [11], the intensity of the colour is not proportional to the concentration of DNA in stained nuclei [12]. Moreover, nuclei are adequately stained by haemalum solutions after prior extraction of all nucleic acids from the sections [13,14]. There is evidence that the Al^{3+}–haematein complex is bound by both ionic forces and van der Waals attraction. The glycerol (or similar highly polar ingredient of the stain) may contribute to the specificity by forming hydrogen bonds with hydrophilic components of the tissue, excluding the colorant from materials other than nuclear chromatin [3,4]. The staining of nuclei with alum–haematein should not be called basophilia because it does not resemble staining by a cationic (basic) dye. Some strongly basophilic materials, including cartilage matrix, some types of mucus and mast cell granules, are not stained by alum–haematein.

In the second stage of the method a red anionic dye, eosin Y, is used to stain everything in the tissue. It is applied from an acidified solution, which promotes protonation of the amino groups of proteins, to which the dye ions are attracted. The intensity of this counterstain can be adjusted to suit the preferences of the person who will examine the stained preparations.

The H&E method is used more than any other staining method. It is the technique applied routinely to all sections in histopathology laboratories, being supplemented by other 'special stains' and immunohistochemical procedures when necessary. All textbooks and atlases of normal and abnormal histology contain photographs of sections stained by this method. Unfortunately this is not a particularly simple procedure to carry out because it is easy to over- or under-stain with either the haemalum or the eosin. Automated machines that mass-produce H&E sections in large laboratories do not allow human judgement of the staining intensities and they are, therefore, likely to provide sections that are less informative than those stained manually as described here.

Method 2. Staining with H&E

Materials

- 95% Alcohol
- 100% Alcohol
- Xylene
- Saturated aqueous lithium carbonate or solid calcium hydroxide
- Ehrlich's haematoxylin
  This is one of the oldest [14a], most reliable and easiest to use of the many alum–haematein formulations
  - □ 200 ml of water
  - □ 200 ml of 100% ethanol
  - □ 200 ml of glycerol
  - □ 10 ml of glacial acetic acid
  - □ 2 g of haematoxylin (CI 75290)
  - □ 30 g of aluminium potassium sulphate [alum; $AlK(SO_4)_2 \cdot 12H_2O$)

  There should be some undissolved alum in the bottom of the bottle. The original mixture, above, takes about 1 month to ripen, in a loosely capped bottle. The solution is ready to use when it is dark red. It can then be kept and used repeatedly for 5 or more years, as haematein is slowly generated by atmospheric oxidation of the haematoxylin. If the solution is required more urgently, 200 mg of sodium iodate can be added. This will immediately oxidize approximately half the haematoxylin, eliminating the month-long wait for atmospheric ripening.
- Eosin Y (0.5%)
  - □ 2.5 g of eosin Y (eosin yellowish; CI 45380)
  - □ 500 ml of water
  - □ 0.5 ml of glacial acetic acid

  This can be stored for several months and may be used repeatedly. Moulds often grow in this solution, so is should be filtered before use. An eosin solution must not be too acidic or the insoluble red-dye–acid complex will be precipitated.
- Acid–alcohol
  - □ 140 ml of 95% or 100% ethanol
  - □ 1 ml of concentrated hydrochloric acid

- Water, to make the volume up to 200 ml
- Mix before using and use only once.

Procedure

The fixative is not critical, but the effects of inadequate fixation (such as insufficient time in formaldehyde) are made quite conspicuous by the H&E method. The technique is described for paraffin sections, but is also suitable for mounted frozen sections.

1. Dewax the sections and bring them through graded alcohols into water.
2. Stain in Ehrlich's haematoxylin for 1–15 min. The optimum time is usually 2–5 min.
3. Wash in running tap water for 2–3 min. If the sections do not turn blue, make the water slightly alkaline by adding a drop of saturated aqueous lithium carbonate or a pinch of calcium hydroxide.

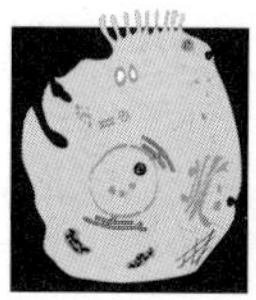

4. Examine the wet section with a microscope. Cell nuclei have strong blue colour. If the sections are understained, return them to Ehrlich's haematoxylin for another 5 min. If they are overstained (blue cytoplasm or collagen), carry out the next step. If staining looks right, omit the differentiation stage (Step 5 and proceed directly to Step 6).
5. *Differentiation*. Agitate the slides in acid–alcohol for 5–10 s, wash in running tap water (or slightly alkaline water) until the sections are blue, and examine again with a microscope. If structures other than the nuclei are stained, return the slides to acid–alcohol. If nuclei seem too pale, return the slides to the dye solution for another 5–10 min. If there is satisfactory nuclear staining proceed with the next step.
6. Stain in eosin for 30–60 s.
7. Wash in running tap water for approx. 30 s. This removes some eosin from the sections. More will be extracted by the alcohols used for dehydration.
8. Dehydrate in 95% alcohol and in two changes of 100% alcohol.

9. Clear in two changes of xylene and apply coverslips, using a resinous mounting medium.

Results

Nuclei stain blue or purple; collagen and the cytoplasm of most cells stain red to pink, depending on the intensity of the eosin counterstain. There should be some variation among different components of the tissue, ideally with an orange cast in the otherwise pink colour of erythrocytes.

3.3. Sulphuric acid–Toluidine Blue

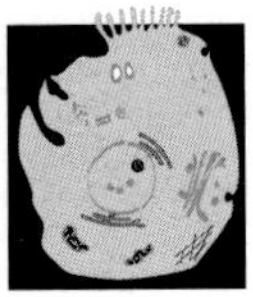

More detail can be resolved in a 0.5–1.0 μm section of a well fixed plastic-embedded specimen than in a paraffin or frozen section of the same object. However, the usual two-colour oversight stains do not work well for various reasons. If the tissue has been fixed in glutaraldehyde, dye-binding can be modified by the introduction of bound but unreacted aldehyde groups. Post-fixation in osmium tetroxide also modifies stainability, by oxidative deamination of proteins and binding of lower osmium oxides to most parts of the tissue. Fortunately, it is usually possible to see considerable structural detail in a thin section coloured by only one dye, if this can be induced to bind to all parts of the sectioned specimen.

Alkaline solutions of Toluidine Blue and other basic dyes are traditionally used for this purpose [8]. At pH 9 or higher all the carboxyl groups of carbohydrates and proteins, including those generated by osmium tetroxide at sites of amino groups, are ionized and can attract basic dye molecules. More anionic sites can be formed by sulphation. This chemical pre-treatment converts hydroxyl and amino groups of organic compounds to half-sulphate esters, which are always ionized and are able to attract dye cations.

Some plastic embedding media, notably epoxy resins and glycol methacrylate, which are cross-linked and cannot be completely removed, impede the diffusion of dyes. This effect can be lessened by etching with a solution containing ethoxide ions. These react with and break some of the bonds that cross-link the resin, thereby increasing its porosity.

Method 3. Staining with sulphuric acid–Toluidine Blue

Materials

- Glass rod or Pasteur pipette
- Distilled water
- 100% Alcohol
- Xylene
- Resinous mounting medium
- Etching solution
  - 1 g of sodium hydroxide or potassium hydroxide
  - 50 ml of 100% ethanol

  Store in a tightly capped bottle. Replace when it becomes brown (from impurities leached from the bottle cap).
- Sulphation reagent
  - Concentrated sulphuric acid (96%), used as described below.
- Dye solution
  - 1 g of Toluidine Blue O (CI 52040)
  - 1 g of borax (sodium tetraborate, $Na_2B_4O_7 \cdot 10H_2O$)
  - 200 ml of water

  This solution can be stored indefinitely. Alternatively, use any of the cationic dye solutions recommended as counterstains later in this chapter.

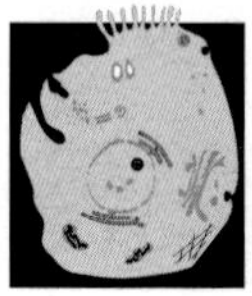

Procedure

This method is for semi-thin sections of plastic-embedded tissue that have been dried down onto glass slides. The first two steps are optional and they need only be performed if there is inadequate staining without etching.

1. Immerse the slide in the etching solution for 10 min.

2. Rinse in two changes of water and air-dry.

3. Using a glass rod or a Pasteur pipette, place a small drop of concentrated sulphuric acid on the sections. Wait for 10 s, then rinse the slide in tap water.

Caution Wear eye protection. Do not get sulphuric acid on your skin or clothes.

4. Immerse in the dye solution for 1 min.
5. Rinse in tap water until no more colour is extracted, then rinse in distilled water.

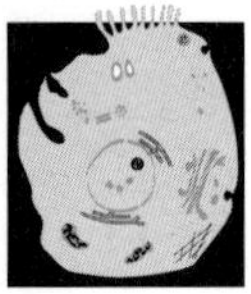

6. Either air-dry thoroughly or dehydrate in two changes of 100% alcohol and clear in xylene.
7. Air-dried sections may be examined without coverslips, using an oil immersion objective. For a permanent preparation apply a coverslip, using a resinous mounting medium. For optimal transparency, mount in the same resin that was used for embedding, and let it polymerize under the coverslip at 60 °C.

Result

Components of the tissue are blue, purple and sometimes red. The latter two colours are due to metachromatic staining, seen when closely adjacent dye molecules, attached to half-sulphate ester groups, cause a shift in absorption to lower wavelengths.

4. Methods that emphasize extracellular materials

These techniques impart strong colours to extracellular structural proteins or carbohydrates. In animals they provide sharp contrast between collagen fibres and the cytoplasm of elongated cells such as muscle fibres. The method for plants gives different colours to cellulose and lignin. Nuclei are also stained in these methods.

4.1. van Gieson's method

In this method, a nuclear stain is followed by a mixture of two anionic dyes at low pH. In Weigert's iron haematoxylin, ferric ions oxidize haematoxylin to haematein and then combine with the haematein to form a blue–black dye–metal complex. Applied from an acidified solution, this stains nuclei selectively, perhaps in the same way as the haemalum solution used in the H&E method. The blue nuclear stain imparted by a haemalum would be extracted by van Gieson's acidic mixture of anionic dyes, but the iron–haematein complex is more resistant.

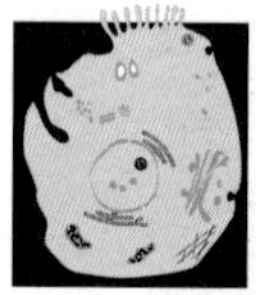

There have been several investigations of differential coloration of cytoplasm and collagen by van Gieson's mixture of picric acid (yellow) with acid fuchsine (red). The results have led to conflicting ideas about the staining mechanism [4]. It seems likely that as they diffuse into the tissue, picrate ions compete for binding sites with the larger and, therefore, more slowly diffusing anions of acid fuchsine [1–3]. The proteinaceous matrix of fixed cytoplasm impedes diffusion of the larger dye ions, so that picrate ions are the first to occupy binding sites there, with van der Waals forces probably holding them in contact with hydrophobic domains of protein molecules [15]. The spaces around and within collagen fibres are assumed to be wider than those within cytoplasm, so that both dyes penetrate at approximately the same rate. The binding of acid fuchsine involves both non-ionic forces and electrostatic attraction to the basic side-chains of collagen [15–17] and, consequently, the attached dye ions are not easily displaced by the smaller picrate ions.

After staining, the sections are washed in acidified water; a low pH prevents loss of bound anionic dyes. Dehydration is carried out in 100% alcohol because most dyes can be extracted by water/alcohol mixtures.

Method 4. Staining by van Gieson's method

Materials

- Weigert's iron haematoxylin
  The working solution is prepared from two stock solutions, A and B.

  Solution A
  - □ 5 g of haematoxylin (CI 75290)
  - □ 500 ml of 95% ethanol

  This is stable for approx. 2 years if the solvent does not evaporate.

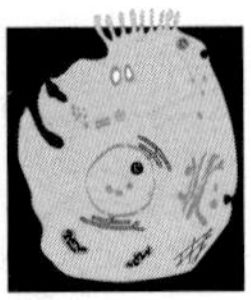

  Solution B
  - □ 5.8 g of ferric chloride ($FeCl_3 \cdot 6H_2O$)
  - □ 495 ml of water
  - □ 5 ml of concentrated hydrochloric acid

  The solution is stable for at least 4 years. If there is a precipitate in an old solution, or if the colour has changed from yellow to orange, add 1 ml of concentrated hydrochloric acid. If this does not provide a clear yellow liquid, make a new batch.

  Working solution
  - □ 50 ml of solution A
  - □ 50 ml of solution B

  Pour solution B into solution A for more rapid mixing. This mixture keeps for approx. 3 days at room temperature, or for 2 weeks at 4 °C, and gradually loses potency with time.
- Acid–alcohol
  - □ 140 ml of 95% or 100% ethanol
  - □ 1 ml of concentrated hydrochloric acid
  - □ Water, to make the volume up to 200 ml

  Mix before using and use only once.
- Van Gieson's stain
  - □ 0.5 g of acid fuchsine (CI 42685)

- □ 500 ml of a saturated aqueous picric acid solution

The composition of this solution may need to be adjusted, depending on results obtained. If sections are too red (e.g. if muscle stains orange when it should be yellow), add 2.5 ml of concentrated hydrochloric acid to the solution. If the red is too weak (pink collagen), add another 0.1 g of acid fuchsine. Once it is working properly this solution can be kept and used repeatedly for at least 5 years.

- ■ 100% ethanol
- ■ Xylene
- ■ Staining rack or a Coplin jar

Procedure

This method works well after any fixation and it can be applied to hydrated paraffin sections or to frozen sections mounted on slides.

1. Stain nuclei in the working Weigert's iron–haematoxylin solution for 3 min. Rinse in tap water and check a wet slide under a microscope. Nuclei should be black and there should be little or no grey colour in other components of the tissue.

2. This step may not be necessary. If specimens are overstained, i.e. if the colour is not confined to nuclei, differentiate in acid–alcohol (5–10 s, with agitation, followed by 15–30 s in running tap water). Check under the microscope, and repeat the differentiation if necessary until only the nuclei are black. For later slides, add some more ferric chloride (solution B) to the working staining solution. If they are understained (nuclei appear grey), return to the iron–haematoxylin mixture for another 3 min. If nuclei are still too pale, increase the proportion of haematoxylin (solution A) in the working mixture.

3. Wash in running tap water for 1 min.

4. Immerse slides in van Gieson's stain for 2 min.

5. Wash in two changes of slightly acidified water to remove excess dye, shake the slides and transfer them to a clean, dry staining rack or Coplin jar.

6 Dehydrate in three changes of 100% alcohol.

7. Clear in xylene and apply coverslips, using a resinous mounting medium.

Results

Nuclei stain black; collagen stains red, but basement membranes and the thinnest fibres are usually unstained; cytoplasm, keratin and erythrocytes stain yellow.

Prolonged immersion in the second change of alcohol (Step 6) will slowly extract picric acid, but not acid fuchsine, from the stained sections. This may be desirable if van Gieson's stain is applied to sections that already contain an insoluble coloured substance such as the brown oxidation product of diaminobenzidine (DAB; 3,3′-diaminobenzidine tetrahydrochloride), which is much used in immunohistochemistry.

4.2. Gabe's one-step trichrome method

A trichrome method is one in which three anionic dyes, with small, medium-sized and large ions, are applied to sections in conjunction with phosphomolybdic acid (PMA; also known as molybdophosphoric acid) or phosphotungstic acid (PTA). PMA and PTA are inorganic compounds with ions that are comparable in size with large dye molecules. Baker [1] called them 'colourless acid dyes'. The pattern of differential coloration is usually attributed to competition among these anions as they penetrate and bind to tissue [1–3], but the nature of dye binding is not yet properly understood [4]. The largest dye ions stain collagen (as with van Gieson's mixture), cartilage and some types of mucus and secretory granules. PTA and PMA also enter collagen and have the effect of excluding the smaller and medium-sized dye anions. The smallest dye ions are the only ones to enter and bind to the contents of erythrocytes and some types of secretory granule. The medium-sized dye ions stain other cytoplasms, and also nuclei if these have not previously been stained.

In the classical trichrome methods of Mallory, Masson and Heidenhain, mixtures of dyes and PMA or PTA are applied sequentially, which necessitates skill and practice but provides some control over the staining, especially in the cytoplasms of different cell types. These procedures all produce strong, dark colours and are therefore suitable only for thin (ideally 4 μm) sections. There are technically simpler techniques in which all three dyes and the PMA or PTA are combined in a single solution. Of these, only one is suitable for sections of any thickness up to about 12 μm [6].

Method 5. Gabe's one-step trichrome method

Materials

- Wiegert's iron–haematoxylin (see Method 4)
- Gabe's trichrome stain
  - 2.5 g of amaranth (CI 16185)
  - 2.5 g of PMA
  - 1.0 g of Fast Green FCF (CI 42053)
  - 500 ml of water
  - 5 ml of glacial acetic acid

  Stir until all the ingredients are dissolved (15 min on a magnetic stirrer), then add either:
  - 0.5 g of Martius Yellow (CI 10315), or
  - 0.05 g of Naphthol Yellow S (CI 10316)
- 100% Alcohol
- 70% Alcohol
- Xylene
- 0.5% Iodine (in 70% alcohol) (only if Steps 2 and 3 are used)
- 5% Aqueous sodium thiosulphate ($Na_2S_2O_3 \cdot 5H_2O$) (only if Steps 2 and 3 are used)
- Acid–alcohol (see Method 2)
- Slightly acidified water
- Resinous mounting medium

  Leave on the magnetic stirrer for approx. 60 min. Filter to remove undissolved yellow dye (if Martius Yellow is used). The solution is stable

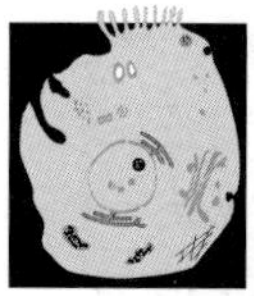

for 5 years, even with frequent reuse. Filtration is needed only if bits of detached sections have found their way into the stock bottle.

As with all trichromes, this method will work after many aqueous fixative mixtures, including buffered formaldehyde, but the colours will be less bright than after a fixative containing mercuric chloride ($HgCl_2$). Bright colours can also be obtained after Bouin's fixative.

Procedure

If the fixative did not contain $HgCl_2$ hydrated sections can be immersed in saturated (approx. 7.5%, w/v) aqueous $HgCl_2$, overnight at room temperature. The solution keeps indefinitely and may be reused many times.

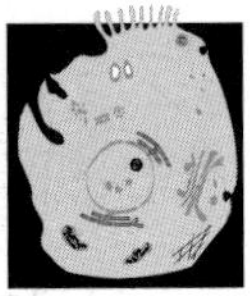

Caution $HgCl_2$ is poisonous; used solutions should not be put down the drain.

This is a method for paraffin sections, which may be up to 12 μm thick.

1. Dewax sections and bring to 70% alcohol.

   The next two steps are necessary if the fixative contained mercuric chloride or if the sections were treated with mercuric chloride as described above. Otherwise Steps 2 and 3 may be skipped.

2. Remove mercury deposits by immersion in 0.5% iodine (in 70% alcohol) for 3 min.
3. Remove the iodine stain by immersion in 5% aqueous sodium thiosulphate solution until sections are no longer yellow or brown (approx. 15 s).
4. Wash in tap water (three changes, or running water for 1 min), and then in distilled water for 30–60 s. If the fixative did not contain mercuric chloride, and the iodine–thiosulphate treatment was omitted, simply place in distilled water for 1 min with agitation for the first 30 s.

5. This step is optional but recommended. Stain in Weigert's iron–haematoxylin for approx. 3 min. Differentiate in acid–alcohol to obtain a pure nuclear stain, and wash for 3 min in running tap water. For a more detailed account of how to stain cell nuclei with iron–haematoxylin, see Method 4.
6. Immerse the slides in Gabe's trichrome stain for at least 10 min. Longer times, at least up to 30 min, can be used.
7. Rinse with agitation in two changes of slightly acidified water, for approx. 30 s each, to remove unbound dyes. Longer times in acidified water do no harm.
8. Dehydrate in three changes of 100% alcohol, clear in xylene and apply coverslips, using a resinous mounting medium.

Results

Nuclei appear black if stained with iron–haematoxylin, but are red otherwise; the cytoplasm stains pink, red or purple (usually with a grey cast after iron–haematoxylin); erythrocytes should be yellow ideally, but are sometimes pink; collagen, mucus and cartilage matrix stain bluish green.

Prior staining of nuclei with iron–haematoxylin reduces the brightness of the trichrome colours but increases the amount of structural detail visible in the sections.

4.3. Orcein method for staining elastin

Elastin is a hydrophobic protein that resists extraction by acids, alkalis and most proteolytic enzymes. It occurs as corrugated laminae in the walls of arteries and as extracellular fibres in several organs, including the skin and lung, in the cartilage of the external ear, and in some ligaments. Multiple elastic laminae account for most of the thickness of the walls of large arteries such as the aorta. The single laminae of small arteries become multiple in response to increased arterial pressure.

The peptide chains of elastin are covalently connected by desmosine and isodesmosine cross-links, which are substituted pyridinium ions formed by the union and cyclization of oxidized and unaltered side chains of the amino acid lysine. Within the amorphous matrix of an elastic fibre or lamina are 11–12 nm microfilaments, composed of a glycoprotein, which constitute approx. 10% of the material [18]. Elastin of immature animals contains free aldehyde groups that are stainable with Schiff's reagent. These are probably oxidized lysine side chains that have not yet reacted to form desmosine or isodesmosine cross-links [19].

Elastic tissue staining is useful for outlining tubes and other cavities in a variety of situations, and for showing pathological changes in blood vessels. Orcein is the easiest dye to use for staining elastic fibres and laminae. There are, however, other methods that provide darker coloration and more stark contrast [4,7–9].

Orcein, is a synthetic product that is made by oxidizing phenolic compounds that were formerly extracted from lichens. It is a mixture of a dozen oxazine dyes with large molecules that bear several methyl groups as side chains [20]. It is applied from an acidic solution, in which the dye exists largely as uncharged molecules. The staining mechanism has not been thoroughly studied. Other dyes that bind to elastin are believed to do so by forming non-covalent bonds (but not hydrogen bonds) between closely approximated hydrophobic parts of the elastin and dye molecules [21].

Orcein may be used alone, or it may precede or follow almost any stain that is not brown or red. Here the prescribed counterstain is a simple blue basic dye.

Method 6. Orcein method for staining elastin

Materials

- Orcein solution
  - 1 g of orcein (CI Natural Red 28)
  - 100 ml of 70% alcohol

Heat to 60 °C with stirring. Let the solution cool to room temperature. Add 1 ml of concentrated hydrochloric acid. Filter to remove any undissolved material. The solution keeps for approx. 1 year and may be re-used many times. It should be warmed to 37 °C in a covered staining vessel before using.

- Toluidine Blue (counterstain)
  - 1 g of Toluidine Blue O (CI 52040)
  - 200 ml of water
  - 2 ml of glacial acetic acid

  The pH should be approx. 4. The solution may be made up in a 0.05 or 0.1 M acetate buffer for more accurate control of the pH. It can be stored for at least 5 years and may be used repeatedly. Filter before use, and check the pH from time to time.
- Graded alcohols
- Xylene

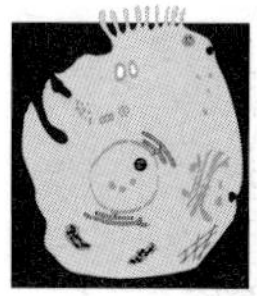

Procedure

This method works after any fixative. The method is described for use with paraffin sections. Mounted frozen sections should be immersed for a few minutes in 100% alcohol before proceeding to Step 2 of the procedure.

1. Dewax the sections and bring them through graded alcohols into water.
2. Agitate the slides in 70% alcohol for 30–60 s.
3. Put the slides in the pre-warmed (37 °C) orcein solution for 45 min.
4. Rinse a slide in 70% alcohol and examine the wet section with a microscope. Elastic laminae and fibres should be strongly stained (brown), and there will probably also be brown or purple coloration of nuclei and collagen.
5. If elastin staining (most easily seen in walls of arteries) seems inadequate, return the slide to the orcein solution for another 15 min and then check again. If the total time in orcein exceeds 90 min the solution is too old or the dye is of no use.

If the slides are to be counterstained (optional), the next step (differentiation) will be necessary. If you do not intend to counterstain the sections, proceed directly to Step 7 (washing).

6. *Differentiation*. Handle slides individually. Move the slide back and forth in acid–alcohol for 2 or 3 s. Wash in tap water, and check with a microscope for retention of brown colour in elastic laminae and fibres. This may have to be repeated. If you extract too much colour from the elastin, return the washed slides to the orcein solution for 15–30 min.

7. *Washing*. Wash the slides in running tap water for at least 30 s. A prolonged immersion in water does not damage the specimen.

8. *Counterstaining*. A blue or green cationic dye is usually satisfactory. Toluidine Blue at pH 4 is recommended. If the sections are somewhat overstained it will be possible to see many structures other than nuclei.

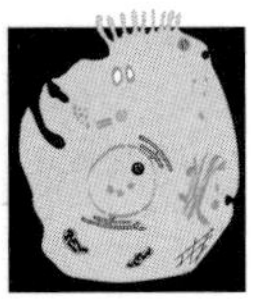

9. Dehydrate in 95% alcohol and two changes of 100% alcohol.

10. Clear in two changes of xylene and apply coverslips, using a resinous mounting medium.

Results

Elastic fibres and laminae stain brown; with the Toluidine Blue counterstain, other tissue components appear in various shades of blue and purple.

4.4. Safranine and Fast Green FCF

In a traditional method for botanical specimens, staining with a cationic dye (safranine) is followed with an anionic dye (originally Light Green SF). In this variant of the traditional technique [22] the anionic dye is Fast Green FCF, which has almost the same colour as Light Green but is less prone to fading during storage of the stained preparations. The cationic dye attaches to nuclei (phosphate anions of DNA) and other structures that contain organic acids (cuticle, lignified cell walls, cartilage matrix, some plastids and other intracytoplasmic bodies). The anionic dye stains everything else.

Method 7. Staining with safranine and Fast Green FCF

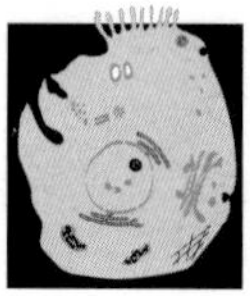

Materials

- Aqueous safranine
  - 2 g of safranine O (CI 50240)
  - 195 ml of water
  - 5 ml of methanol

  This solution may be kept and reused for at least 2 years. Filter if necessary before using. The methanol suppresses the growth of micro-organisms.
- Alcoholic Fast Green FCF
  - 1 g of Fast Green FCF (CI 42053)
  - 200 ml of 95% alcohol

  This solution is stable and may be reused frequently for at least 5 years. Old solutions may need to be filtered.
- 30%, 50%, 70% 95% and 100% Alcohol
- Distilled water
- Xylene
- Resinous mounting medium

Procedure

Pieces of plants are commonly fixed in alcohol–acetic acid mixtures, with or without formaldehyde. Fixation is not critical, however, for this simple technique. The method is described for paraffin sections, but it should also be suitable for mounted frozen or cryostat sections. The procedure includes a differentiation that must be checked visually; therefore handle the slides individually through Steps 4–7 unless you have already determined the optimal timing for routine use with your materials. The principle of the differentiation (Step 6) is that safranine is extracted by alcohol but not by water. Fast Green FCF can be removed by water (especially neutral or slightly alkaline tap water). Absolute alcohol is used for the final dehydration (Step 7) because both dyes are susceptible to extraction by alcohol–water mixtures.

1. Dewax and hydrate paraffin sections.
2. Immerse slides in aqueous safranine for 1 h. Longer immersion does not damage the section; there should be considerable overstaining.
3. Wash in running tap water until no more red dye comes out of the sections, then rinse in distilled water.
4. Pass slide through 30%, 50%, 70% and 95% alcohol for approx. 30 s each with agitation. Some safranine is extracted into these alcohols.
5. Stain in alcoholic Fast Green FCF for 5–10 s.

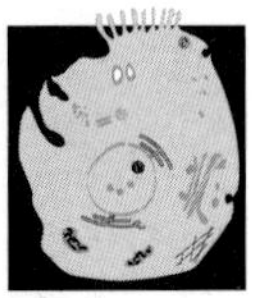

6. Rinse in 100% alcohol and check under a microscope. The red colour should be seen only in cuticle, lignified cell walls, nuclei and some plastids (nuclei, cartilage matrix and some types of mucus in animal tissues). Repeat Step 5 to remove more safranine if necessary. The slide can remain in the Fast Green solution for as long as 2 min. If too much safranine has been removed, wash the slide in running tap water until no more green remains in the sections, and return to Step 2.
7. Dehydrate in three changes of 100% alcohol (20–30 s in each) with agitation.
8. Clear in two changes of xylene and cover, using a resinous mounting medium.

Result

Lignified cell walls, cuticles and nuclei appear red; cellulose cell walls stain green; plastids (including chloroplasts) may be red or green. In animal tissues nuclei, cartilage matrix, some mucus and some cytoplasmic granules are red; collagen, keratin and most cytoplasms are green.

5. Methods for staining nervous system tissue

The anatomy of the vertebrate central nervous system is defined largely in terms of the distribution of neurons and bundles of myelinated axons. Unfortunately, staining methods useful with non-neural tissues do not adequately show the neuronal cell bodies in grey matter or the myelin sheaths of axons, which are the most prominent components of white matter. Staining for myelin provides the easiest way to identify structures in the cerebral hemispheres. A Nissl stain, which colours all cell nuclei and neuronal cytoplasmic RNA, is useful for identifying groups of neuronal cell bodies by their sizes, shapes, and intracytoplasmic arrangements of RNA-rich rough endoplasmic reticulum (Nissl bodies).

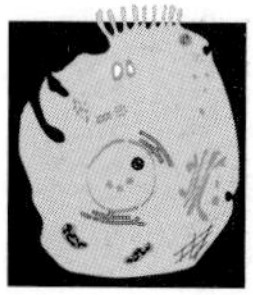

One myelin/Nissl method is described here. Traditional methods for axons and neuroglial cells are unlikely to be used in association with the techniques of molecular biology.

5.1. Iron–eriochrome cyanine R and Neutral Red

In the presence of an excess of ferric ions, eriochrome cyanine R is present in anionic dye–metal complexes [23] that attach to all parts of the section. Treatment with either a weak alkali or a solution of ferric chloride extracts the colour from nuclei and cytoplasm more quickly than from myelin and erythrocytes. The mechanisms of staining and differentiation are not understood. Myelin staining is followed by application of a red cationic dye, which is attracted to nucleic acids. The pH of the dye solution is low enough to suppress the ionization of protein carboxyl groups, leaving the phosphates of DNA and RNA as the only macromolecular anions present in large quantity in nervous tissue.

This method [9b, 24] is highly recommended because it takes less than 1 h and imparts strong colour. Other basic dyes may be used instead of Neutral Red, or the Nissl stain may be omitted.

Method 8. Staining nervous tissue with iron–eriochrome cyanine R and Neutral Red

Materials

- Iron–eriochrome cyanine R solution for myelin
  - 20 ml of 0.21 M ferric chloride.
  - 1.0 g of eriochrome cyanine R (CI 43820)
  - 2.5 ml of concentrated (96%) sulphuric acid
  - Water to make volume up to 500 ml

  This solution can be used repeatedly for 10 years.

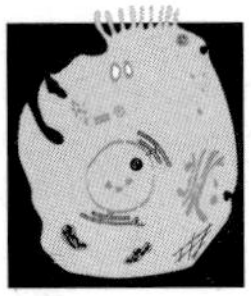

- Differentiating solution
  - 16.8 g of ferric chloride ($FeCl_3 \cdot 6H_2O$)
  - 250 ml of water

  When dissolved, add more water to make the volume up to 300 ml. This keeps for approx. 3 years. Replace it if there is a precipitate at the bottom of the bottle.
- 0.5% Neutral Red solution
  - 2.0 g of Neutral Red (CI 50040)
  - 400 ml of water
  - 2 ml of glacial acetic acid

  The pH should be in the range 3.5–4.0. From time to time, filter and adjust the pH by adding drops of 1.0 M acetic acid (5.7%, v/v) or 1.0 M sodium hydroxide (4%, w/v). This dye solution keeps for at least 10 years and may be used repeatedly.
- Graded alcohols
- Xylene
- Resinous mounting medium

Procedure

This method is usually carried out on paraffin sections of specimens of central nervous tissue that have been fixed in neutral buffered formaldehyde for at least 2 days (preferably longer). Most other aqueous fixative solutions are also suitable. Frozen sections, dried onto slides, may also be used; for thick sections the times need to be longer than those prescribed.

To display myelin alone, omit Steps 6, 7 and 9 of the complete procedure. To display only the Nissl substance and nuclei, omit Steps 2–5.

1. Dewax the sections and bring them through graded alcohols into water.
2. Immerse in the iron–eriochrome cyanine R solution for at least 10 min. This time can be extended up to 30 min.
3. Wash in running tap water to remove excess dye.
4. Immerse in the differentiating solution, with gentle agitation, until areas of grey matter in the section are conspicuously paler than areas of white matter. Nuclei of cells should not be stained.

   The differentiation can be stopped by moving the slides into running tap water while you check a section. Typically, the total time needed in the ferric chloride differentiator is approx. 5 min, but with some tissues it may be as short as 30 s or as long as 10 min.

5. Wash in running tap water for 1 min.
6. Stain in 0.5% Neutral Red for 1 min.
7. Wash in running tap water for 30–60 s.
8. Dehydrate in 95% alcohol and in two changes of 100% alcohol.
9. Check under a microscope. Red neurons should be clearly visible in grey matter.

   If the red colour is inadequate, rinse in water and repeat Step 6. If there is too much red (look in the neuropil between neuronal cell bodies), immerse the slide in 70% alcohol to remove some Neutral Red, then repeat Step 8.

10. Clear in two changes of xylene and apply coverslips, using a resinous mounting medium.

Results

Myelin sheaths of nerve fibres (and also erythrocytes) stain dark blue; nuclei and Nissl substance stain red.

6. Counterstains

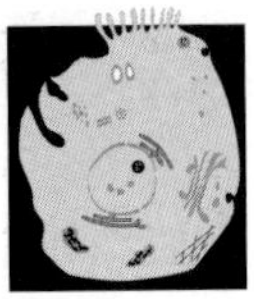

After carrying out some specific histochemical procedure (such as *in situ* hybridization or an immunohistochemical method) a counterstain is often applied to display the architecture of the tissue or to show the distribution of different cell types. Usually a counterstain should have a colour that contrasts with the principal stain but is paler (Table 1). A counterstain allows the investigator to recognize the sites of the reaction product of the principal procedure. A simple pink, blue or green stain for nuclear chromatin may be all that is needed. If cytoplasm and extracellular materials are to be made visible, they are usually rendered in pink, yellow or pale green.

It is easy to wreck a preparation by not thinking before counterstaining. Before applying a counterstain you must ask yourself three questions.

1. What colour should the counterstain be?

   The answer depends on the colour of the product already present in the preparations (see Table 1).

2. What dissolves the reaction product of the principal staining method?

   The answer might be 'nothing' or 'alcohols' or "xylene'. A counterstaining procedure must not include passage through a solvent that would dissolve the most important fruit of your endeavour. In general, polymeric reaction products (such as the product of DAB oxidation) are not extracted by anything. The products of indigogenic methods of enzyme histochemistry are similarly insoluble, and so are the diformazans of Nitro-Blue Tetrazolium and Tetranitro-Blue Tetrazolium. Other products of enzyme histochemical reactions, including most azo dyes, the oxidation product of 9-aminoethylcarbazole and some formazans, are soluble in alcohol and other organic solvents. Inorganic products are insoluble in organic solvents but may be decolourized by reaction with acids or bases or by oxidation or reduction during storage of the mounted preparations.

3. What would extract the counterstain?

   Counterstains are dyes, and are susceptible to extraction from their substrates by water, acidic or alkaline solutions or alcohols. In general, anionic and cationic dyes are extracted by alkaline and acidic solutions respectively. Both are likely to be removed slowly by plain water or an aqueous mounting medium. Mordant dyes used for staining nuclei are extracted slowly by acids, but are resistant to most of the other reagents that might be used in processing stained slides.

Table 1. Dyes used as counterstains

Area/material to be viewed	Type of dye needed	Example(s)
Nuclei only	Mordant dyes with aluminium	Alum–tetracycline (yellow fluorescence); haemalum (blue); nuclear Fast Red (pink)
	DNA-intercalating fluorochromes	Bisbenzimide (Hoechst 33258) (blue fluorescence)
Nuclear DNA and other polyanions (RNA, mucus, mast cells, cartilage, lignin)	Cationic (basic) dyes	Toluidine Blue (blue); Ethyl Green (greenish blue); Neutral Red (red; orange fluorescence); Acridine Orange [fluorescent; DNA (green), RNA (red)]
Cytoplasm and extracellular materials	Anionic (acid) dyes	Eosin (pink to red; green fluorescence); Fast Green FCF (green); Metanil Yellow (yellow)

Table 2. Properties of products of histochemical reactions and of dyes used as counterstains

Name	Uses and examples
Visible organic end-products of enzyme histochemistry	
Most azo dyes	Hydrolytic enzyme activities, including esterases and alkaline phosphatase
Azo dyes from hexazonium pararosaniline	Hydrolytic enzyme activities, including esterases and alkaline phosphatase
Bromoindigos	Methods for phosphatases, esterases and glycosidases with 5-bromoindoxyl or 4-chloro-5-bromoindolyl derivatives as substrates
Blue diformazan of Nitro-Blue Tetrazolium, & brown diformazan of Tetranitro-Blue Tetrazolium	Methods for dehydrogenases and some other enzymes, using Nitro-Blue Tetrazolium or Tetranitro-Blue Tetrazolium salts in the incubation medium
Most other formazans (red or blue)	Methods for dehydrogenases; also methods using tetrazolium salts to detect living cells or tissues
Melanin	Methods for DOPA oxidase (phenol oxidase)
Oxidized 9-aminoethylcarbazole	Peroxidase methods using 3-amino-9-ethylcarbazole as chromogen
Oxidized DAB	Peroxidase methods using 3,3'-diaminobenzidine as chromogen; methods for oxidases and other enzymes
Inorganic end-products of enzyme histochemistry and other methods	
Cobalt sulphide	Calcium/cobalt precipitation methods for ATPase and other alkaline phosphatases
Copper ferrocyanide (Hatchett's brown)	Karnovsky–Roots method for choline esterases
Copper sulphide	Koelle's method for choline esterases
Ferric ferrocyanide; ferrous ferricyanide (Prussian Blue, Turnbull's Blue)	Perls' method for iron; ferric ferricyanide method for reducing groups in tissue
Gold	Many traditional techniques; toning of silver deposits; label for macromolecules, including antibodies
Lead sulphide	Lead precipitation methods for acid phosphatase
Osmium dioxide (from reduction of osmium tetroxide)[2]	Many traditional techniques; lipid histochemistry; myelin of teased nerve fibres
Silver[3]	Many traditional techniques; physical developers used to amplify metal sulphides, DAB product or colloidal gold

Table 2 (contd.)

Dissolved or extracted by: Aqueous acid	Aqueous alkali	Alcohol–water mixtures	100% alcohol	Xylene	Suitable type of mounting medium
No	No	Yes	Yes	Yes	Aqueous
No	No	No	No	No	Resinous
No	No	No	No	No	Resinous
No	No	No	No	No	Resinous[1]
No	No	Yes	Yes	Yes	Aqueous
No	No	No	No	No	Resinous
No	No	Yes	Yes	Yes	Aqueous
No	No	No	No	No	Resinous
Yes	No	No	No	No	Resinous (Canada balsam preferred)
Slightly	No	No	No	No	Resinous
Slightly	No	No	No	No	Resinous
Yes (oxalic) No (others)	Yes	No	No	No	Resinous (synthetic preferred)
No	No	No	No	No	Resinous
Slightly	No	No	No	No	Resinous
No	No	No	No	No	Resinous
Yes (HNO_3) No (others)	No	No	No	No	Resinous

(contd.) ☞

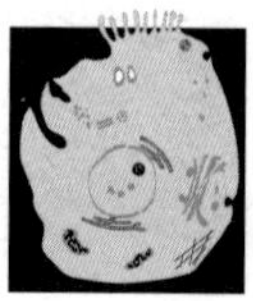

Table 2. (contd.)

Name	Uses and examples
Dyes	
Iron–mordant dye complexes	Staining nuclei black (Weigert's haematoxylin) or blue (iron–eriochrome cyanine R); iron–haematoxylin methods for myelin, organelles and chromosomes
Aluminium–mordant dye complexes	Staining nuclei blue (haemalum, as in H&E)
Cationic (basic) dyes	Staining DNA, RNA, mast cells, cartilage, mucus, lignin and other polyanions
Anionic (acid) dyes	Staining proteins and other background materials
Reactive dyes	Schiff's and related reagents, used in Feulgen reaction for DNA, periodic acid–Schiff methods for carbohydrates, and in some methods for lipids.
Solvent dyes[4]	Staining of lipids (Oil Red O, Sudan Black B, Nile Red)
Fluorescent end-products	
Labelled antibodies[5]	Immunofluorescence methods
Cationic fluorochromes	Intercalating dyes for DNA and many red and yellow basic dyes
Anionic fluorochromes	Eosin and other fluorescent counterstains
Fluorescent metal complexes	Histochemical reagents for metals; fluorescent stains[6]

[1]The red monoformazans are soluble in alcohol or xylene.

[2]To remove black OsO_2, immerse in 1% H_2O_2, then wash for several hours in running tap water.

[3]Remove silver deposits with Farmer's reducer: equal volumes of 10% $K_3Fe(CN)_6$ and 10% $Na_2S_2O_3 \cdot 5H_2O$.

[4]The 'luxol' dyes, often used for myelin, resist organic solvents but are alkali-labile.

[5]Antigen–antibody binding is acid-labile. The covalent antibody–fluorochrome bond is stable.

[6]General guidelines only. Stability and solubility vary for different metal-fluorochrome stains.

Table 2 (contd.)

Dissolved or extracted by:					Suitable type of mounting medium
Aqueous acid	Aqueous alkali	Alcohol–water mixtures	100% alcohol	Xylene	
Slowly	No	No	No	No	Resinous
Yes	No	No	No	No	Resinous
Yes	No	Yes	No	No	Resinous
No	Yes	Yes	No	No	Resinous
No	No	No	No	No	Resinous
No	No	Yes	Yes	Yes	Aqueous
Yes	No	No	No	No	Aqueous (slightly alkaline)
Yes	No	Yes	No	No	Aqueous (slightly alkaline)
No	Yes	Yes	No	No	Resinous
Yes	No	Yes/no	Yes/no	Yes/no	Aqueous (neutral)

It is important to know the nature of the product of the primary procedure before counterstaining, and also to know what would or would not extract bound dye from the stained preparation. This knowledge determines whether the preparations should be mounted in a resinous or an aqueous medium. The properties of some common end-products of enzyme histochemistry, immunohistochemistry and *in situ* hybridization histochemistry and of some commonly used dyes are summarized in Table 2.

6.1. Nuclear counterstains

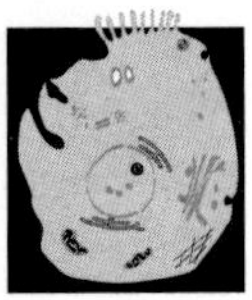

The counterstains selected for this group are three aluminium mordant dyes (yellow-fluorescent, blue and red) and one fluorochrome that binds selectively to DNA.

6.1.1. Haemalum

The staining of nuclei with haemalum has already been described as part of the H&E method. For a blue nuclear counterstain carry out Steps 2–4 of that procedure.

6.1.2. Nuclear Fast Red

This anthraquinone dye, also known as Kernechtrot, forms a red complex with aluminium, which has an affinity for nuclei. It is often impossible to avoid some coloration of cytoplasm and collagen.

Dye solution

- 0.2 g of nuclear Fast Red (CI 60760)
- 10 g of aluminium sulphate [$(Al_2(SO_4)_3 \cdot 18H_2O$]
- 200 ml of water

Heat with stirring until the mixture boils, then leave overnight to cool. Filter to remove insoluble material. The solution can be kept and used repeatedly for approx. 1 year. Filter before using.

Procedure

1. Bring sections to water.
2. Stain in the nuclear Fast Red solution for 5–10 min.
3. Wash in running tap water for approx. 1 min.

6.1.3. Alum–tetracycline

The antibiotic tetracycline and its metal complexes are fluorescent (absorption in near UV and blue; yellow emission). A mixture containing aluminium ions and tetracycline imparts yellow fluorescence selectively to nuclei [4]. This method is not suitable for counterstaining preparations that have been immunostained with fluorescently labelled antibodies because its acidity would break the antigen–antibody bonds.

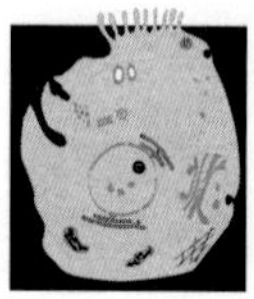

Alum–tetracycline solution

- 0.5 g of tetracycline hydrochloride (the contents of two 250 mg capsules)
- 100 ml of water
- 5.0 g of aluminium potassium sulphate [alum; $AlK(SO_4)_2\cdot 12H_2O$]

This can be kept for approx. 1 month. The colour changes slowly from yellow to brown.

Procedure

1. Bring sections to water.
2. Stain in the alum–tetracycline solution for 5 min.
3. Wash in three changes of water, each for 30 s.
4. Dehydrate, clear and mount in a non-fluorescent resinous medium (media based on polystyrene or polymethylmethacrylate are usually non-fluorescent).

6.1.4. Bisbenzimide (Hoechst dye)

The name bisbenzimide is applied to two of several compounds related to hemicyanine dyes. These are commonly known as Hoechst dyes, each distinguished by a number. Their cations intercalate in adenine- and thymine-rich regions of the minor groove of DNA, and binding greatly increases the fluorescence (UV excitation, blue emission); therefore they stain nuclei. These reagents are commonly used with cultured cells or cryostat sections. The compounds are applied from neutral solutions, so they are suitable for counterstaining preparations that contain fluorescently labelled antibodies. Unfortunately the preparations are not permanent.

Bisbenzimide stock solution

- 10 mg of bisbenzimide (Hoechst 33258; alternatively Hoechst 33342)
- 20 ml of water

This can be kept for at least 2 years at 4 °C.

Working solution

Dilute 1 volume of the stock solution with 500–1000 volumes of any buffer or balanced salt solution (pH 7.0–7.4). The concentration is not critical, but further dilution is sometimes needed if nuclear fluorescence is too bright.

Procedure

1. Fix the cells or sections in alcohol or acetic acid–alcohol (this step is optional because bisbenzimide also enters living cells).
2. Rinse with the same type of buffer or salt solution as that used to dissolve the reagent.
3. Stain for 15–60 min (not critical) in the working solution of bisbenzimide at room temperature.
4. Rinse in three changes of buffer or balanced salt solution.

5. Apply a coverslip, using a mixture of equal volumes of glycerol and buffer (or balanced salt solution) as the mounting medium. Examine and photograph. The slides can be kept for several weeks at 4 °C.

   Alternatively rinse in water and thoroughly air-dry (2 h or longer, at room temperature). Rinse in xylene and apply a coverslip using a non-fluorescent resinous mounting medium. The intensity of the fluorescence is reduced if this is done.

6.2. Cationic counterstains

Four basic dyes are included in this group: blue, green and red for regular light microscopy, and a fluorochrome. In addition to nuclei, these dyes bind to cytoplasmic RNA, some secretory granules, some mucus, and some extracellular materials, notably cartilage matrix.

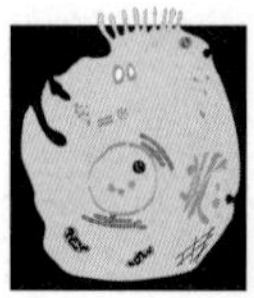

6.2.1. Toluidine Blue

Toluidine Blue solution

For the components see Method 3.

Procedure

1. Stain hydrated sections in 0.5% Toluidine Blue for 1 min.
2. Wash in three changes of water (30–60 s each).
3. Dehydrate in 95% alcohol and two changes of 100% alcohol.
4. Check the staining with a microscope. If it is too weak repeat Step 1. If it is too strong put the slides in 70% alcohol for 15–30 s, then go back to the beginning of Step 3.
5. Clear in two changes of xylene and apply coverslips using a resinous mounting medium.

6.2.2. Ethyl Green (Methyl Green)

For many years dyes sold as Methyl Green (CI 42585, Basic Blue 20) have really been Ethyl Green (CI 43590), which differs only in the replacement of one of its seven methyl groups by ethyl. True Methyl Green (in older stocks of dye) is always contaminated with Crystal Violet, which must be removed by extracting the aqueous solution with several changes of chloroform. Ethyl Green (sold under the name of Methyl Green) is not contaminated with Crystal Violet [25].

Staining solution

- 0.4 g of 'Methyl Green' (Ethyl Green) (CI 42590)
- 0.2 g of potassium hydrogen phthalate ($KHC_8H_4O_4$)
- 200 ml of water

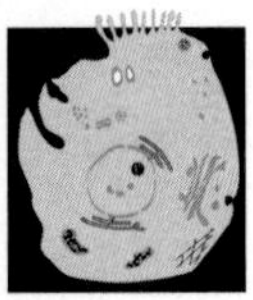

The $KHC_8H_4O_4$ ensures a pH of approx. 4. This solution can be kept for a few months and used repeatedly.

Procedure

1. Stain hydrated sections in the 0.2% Methyl Green solution for 15 min.
2. Rinse quickly in water (which can extract the dye).
3. Blot the sections with filter paper.
4. Air-dry the slides overnight at room temperature, in a dust-free place, or immerse slides in two changes of n-butanol, each for 5 min.
5. Immerse in xylene for 5 min and apply coverslips, using a resinous mounting medium.

6.2.3. Neutral Red

Staining solution

- 2.0 g of Neutral Red (CI 50040)
- 400 ml of water
- 2 ml of glacial acetic acid

The pH should be approx. 4. Filter from time to time and adjust the pH by adding drops of acetic acid or 1.0 M (4%, w/v) sodium hydroxide. The solution lasts for at least 10 years and may be used repeatedly.

Procedure

1. Stain hydrated sections in 0.5% Neutral Red for 1 min.
2. Wash in running tap water for 30–60 s.
3. Dehydrate in 95% alcohol and two changes of 100% alcohol.
4. Check the staining with a microscope; if it is too weak repeat Step 1; if it is too strong put the slides in 70% alcohol for 15–30 s, then go back to the beginning of Step 3.
5. Clear in two changes of xylene and apply coverslips using a resinous mounting medium.

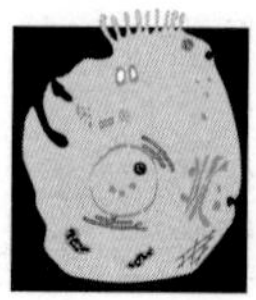

6.2.4. Acridine Orange

The fluorescent emission of this dye is affected by the substrate to which it is bound. The fluorescence of stained DNA is green, whereas that of RNA and other basophilic materials is red. The method works best after an alcoholic fixative. Formaldehyde is detrimental and Bouin's fluid completely prevents the differential fluorescence of the two nucleic acids [26].

Stock solutions

- Acridine Orange (CI 46005) 0.1% in water
- 0.1 M phosphate buffer (pH 6.0)
- Mounting medium (equal volumes of phosphate buffer and glycerol)

These are stable for several months in a refrigerator.

Procedure

1. Dewax and hydrate paraffin sections. Fix cryostat sections or smears of cells in alcohol or Carnoy's fluid for 2 min.

2. Place in phosphate buffer for approx. 1 min while mixing the working staining solution (Step 3).
3. Dilute 0.1% Acridine Orange with 10 volumes of phosphate buffer. This gives a 1:10 000 dye solution. To obtain optimal emission colours it may be necessary to experiment with different dilutions in the range 1:5000 to 1:100 000.
4. Stain in the diluted dye solution for 15 min.
5. Rinse in three changes of phosphate buffer.
6. Apply a coverslip, using the buffered glycerol mounting medium described above.
7. Examine immediately with a fluorescence microscope. The absorption maximum of Acridine Orange is 500 nm (blue), and the emissions of stained DNA and RNA are 530 and 600 nm respectively. The strong fluorescence can, however, be detected with less than ideal filter sets.

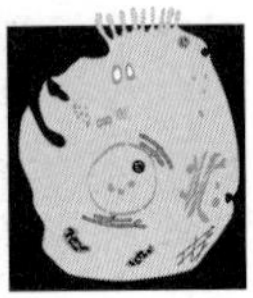

6.3. Anionic counterstains

Nuclear counterstaining is often sufficient, but the architecture of many tissues is shown more clearly by staining collagen and cytoplasm. This amounts to colouring proteins, and anionic dyes provide the simplest way to do this. They are applied from acidified solutions, in which the amino groups of proteins are protonated and are able to attract the negatively charged dye ions. Non-ionic forces are also involved in the binding of these dyes [2–5].

6.3.1. Eosin

This red dye is most frequently used after a blue nuclear stain, but it contrasts well with brown stains (such as the DAB oxidation product) and it is also fluorescent.

Staining solution

- 0.1–0.5% eosin Y (CI 45380) in 0.05–0.1% aqueous acetic acid
  The solution is stable for several months. Replace it if it contains moulds or if the colour is noticeably diminished.

Procedure

1. Immerse the slides in the eosin solution for 30–60 s.
2. Wash in running tap water. This removes some of the dye. For transmitted light microscopy check a wet slide for the desired intensity of staining. For fluorescence microscopy the visible colour should be quite pale.
3. Vigorously shake off the water and dehydrate in three changes of 100% alcohol.
4. Clear in two changes of xylene and apply coverslips using a non-fluorescent resinous mounting medium.

Result

Cytoplasm and collagen appear in various shades of pink. With a fluorescence microscope eosin staining is seen as a green emission similar to, but duller than, that of fluorescein (see Table 1).

6.3.2. Fast Green FCF

Staining solution

- 0.1–0.2% Fast Green FCF (CI 42053) in 0.5–1.0% aqueous acetic acid
  The solution is stable for at least 5 years and can be used repeatedly. Filter it occasionally and replace when staining becomes weak. Light Green SF (CI 42095) may be used instead. Its colour is almost identical but sections stained with Light Green fade with time (years), whereas Fast Green FCF is permanent.

Procedure

1. Immerse the slides in the Fast Green FCF solution for 30–60 s.

2. Wash in running tap water. This removes some of the dye. Check a wet slide for the desired intensity of staining.
3. Vigorously shake off the water and dehydrate in three changes of 100% alcohol.
4. Clear in two changes of xylene and apply coverslips using a resinous mounting medium.

Result

Cytoplasm and collagen stain light green with a bluish cast.

6.3.3. Metanil Yellow

Staining solution

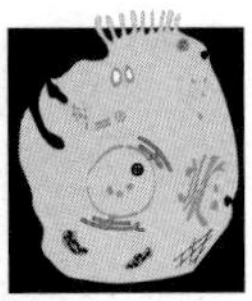

- 0.25–0.5% Metanil Yellow (CI 13065) in 0.25–0.5% aqueous acetic acid. The solution darkens on storage, and its staining power deteriorates. Do not use a solution that is more than 4 weeks old.

Procedure

1. Immerse the slides in the Metanil Yellow solution for 30–60 s.
2. Wash in running tap water. This removes some of the dye. Check a wet slide for the desired intensity of staining.
3. Vigorously shake off the water and dehydrate in three changes of 100% alcohol.
4. Clear in two changes of xylene and apply coverslips using a resinous mounting medium.

Result

Cytoplasm and collagen stain yellow.

7. Removal of coverslips and restaining

Occasionally it is necessary to remove the coverslip from a preparation and restain it. Slides should be soaked in xylene until the coverslip comes off easily. This may take as long as 1 week for an old slide. Soaking in xylene for a further 24 h dissolves residual mounting medium. Slides should then be rinsed in fresh xylene and placed in 100% ethanol. If a translucent or cloudy appearance develops on the slide, mounting medium is still present and another 1–2 days in xylene will be needed to remove it. When the slide is fully transparent in alcohol it may be rehydrated in the usual way.

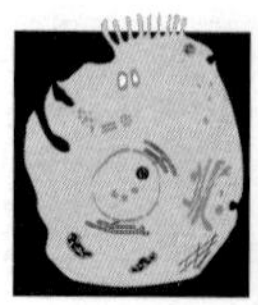

Subsequent treatment depends on the needs of the investigation. To remove stains from the rehydrated sections, immerse in acid–alcohol [70% ethanol with 1% (v/v) concentrated hydrochloric acid]. This extracts most dye–metal complexes and cationic (basic) dyes (Alcian Blue being a notable exception). Anionic (acid) dyes are extracted by immersion in an alkaline solution; saturated aqueous lithium carbonate is often used for this purpose. After removal of dyes the slides are washed for 5 min in running tap water and are then rinsed in distilled water before staining again.

References

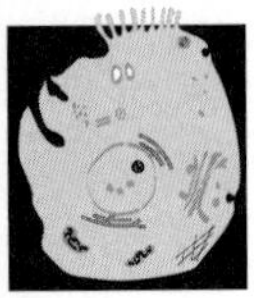

1. Baker, J.R. (1958) Principles of Biological Microtechnique (Reprinted 1970, with corrections), Methuen, London

2. Horobin, R.W. (1982) Histochemistry: an Explanatory Outline of Histochemistry and Biophysical Staining, Gustav Fischer, Stuttgart

3. Horobin, R.W. (1988) Understanding Histochemistry: Selection, Evaluation and Design of Biological Stains, Ellis Horwood, Chichester

4. Kiernan, J.A. (1999) Histological and Histochemical Methods: Theory and Practice, Butterworth–Heinemann, Oxford

5. Lyon, H. (1991) Theory and Strategy in Histochemistry. A Guide to the Selection and Understanding of Techniques, Springer Verlag, Berlin

6. Gabe, M. (1976) Histological Techniques (English edn, translated by Blackith E. and Kavoor, A.), Masson, Paris

7. Culling, C.F.A., Allison, R.T. and Barr, W.T. (1985) Cellular Pathology Technique, Butterworths, London

8. Hayat, M.A. (1993) Stains and Cytochemical Methods, Plenum Press, New York

9. Presnell, J.K. and Schreibman, M.P. (1997) Humason's Animal Tissue Techniques, Johns Hopkins University Press, Baltimore, MD

9a. Hyman, J.M. and Poulding, R.H. (1961) J. Med. Lab. Technol. **18**, 107

9b. Kiernan, J.A. (1984) J. Microsc.**134**, 13–23

10. Marshall, P.N. and Horobin, R.W. (1972) Histochem. J. **4**, 493–503

11. Bettinger, C. and Zimmermann, H.W. (1991) Histochemistry **96**, 215–228

12. Schulte, E.K.W. and Fink, D.K. (1995) Anal. Cell. Pathol. **9**, 257–268

13. Baker, J.R. (1962) Q. J. Microsc. Sci. **103**, 493–517

14. Lillie, R.D., Pizzolato, P. and Donaldson, P.T. (1976) Histochemistry **49**, 23–35

14a. Ehrlich, P. (1886) Z. Wiss. Mikr. **3**, 150

15. Prento, P. (1993) Histochemistry **99**, 163–174

16. Junqueira, L.C.U., Bignolas, G. and Brentani, R.R. (1979) Histochem. J. **11**, 447–455

17. Puchtler, H., Meloan, S.N. and Waldrop, F.S. (1988) Histochemistry **88**, 243–256

18. Chadwick, D.O. and Goode, J.A. (1995) The Molecular Biology and Pathology of Elastic Tissues: Ciba Foundation Symposium 192, Wiley, Chichester

19. MacCallum, D.K. (1973) Stain Technol. **48**, 117–122

20. Lillie, R.D. (1977) H.J. Conn's Biological Stains, Williams & Wilkins, Baltimore, MD

21. Horobin, R.W. and Flemming, L. (1980) J. Microsc. **119**, 357–372

22. Berlyn, G.P. and Miksche, J.P. (1976) Botanical Microtechnique and Cytochemistry, Iowa State University Press, Ames, IA

23. Kiernan, J.A. (1984) J. Microsc. **134**, 13–23

24. Page, K.M. (1965) J. Med. Lab. Technol. **22**, 224–225

25. Green, F.J. (1990) The Sigma–Aldrich Handbook of Stains, Dyes and Indicators, Aldrich Chemical Company, Milwaukee, WI

26. von Bertalanffy, L. and Bickis, I. (1956) J. Histochem. Cytochem. **4**, 481–493

Immuno-histochemical methods

Susan Van Noorden

Chapter 9

Page

1. Introduction

Immunohistochemistry is a way of localizing a substance (antigen) in a tissue preparation by making it react with an antibody raised to that antigen. The antigen may occur naturally (for example, one of the collagen components of connective tissue), or it may be an artificially introduced substance, such as digoxygenin used to label a probe for *in situ* hybridization of nucleic acids. The site of the reaction is identified by attaching a microscopically visible label to the antibody. This chapter provides an introduction to the basic techniques. For wider discussion and further details see references [1–5].

The method exploits the high specificity with which antibodies bind to their antigens. In the first immunohistochemical method described [6], antibodies labelled with a fluorescent compound, fluorescein isothiocyanate (FITC), were used to identify a native immunoglobulin. The method was then extended to the detection of other substances. Enzyme labels were introduced in the 1960s, allowing an ordinary light microscope to be used instead of one with UV excitation. During the 1970s many technical problems were overcome, and immunohistochemistry became a dependable and important adjunct to histological techniques. Immunohistochemistry is valuable in experimental biology and is also now indispensible in diagnostic pathology, especially for determining the origins of the cell types in tumours and in providing information of prognostic significance.

1.1. Uses of immunohistochemistry in molecular biology

In connection with molecular biology, immunohistochemistry is used to identify gene products in transgenic animals, as proof of transfer of the gene, and to localize the sites of an *in situ* hybridization or PCR by identifying the antigenic marker on the applied probe. It is also used in combination with this type of reaction to identify the cell or tissue in which the reaction has taken place. Western blotting uses antigen–antibody reactions in much the same way as immunohistochemistry, except that the substances identified are in tissue extracts after electrophoretic separation. The additional advantage of immunohistochemistry lies in the precision with which antigens can be localized in tissues and cells: to the cytoplasmic or nuclear level with light microscopy and to subcellular levels with electron microscopy. This chapter concentrates on light microscopy. The principles are the same for electron microscopical immunocytochemistry but there are some differences in methodology [5].

1.2. Antibody production: polyclonal and monoclonal antibodies

Many different types of molecule can act as antigens. Immunizing a rabbit (or mouse or other animal) with the substance of interest will stimulate plasma cells (mature B-cells of the immune system) to produce antibodies [immunoglobulins (Igs)] to that substance, which are secreted into the blood. One plasma cell can make only one type of antibody, but after immunization with a large antigen molecule several types of antibody, specific for different immunogenic areas (epitopes) of that molecule, will be produced, each from a different clone of plasma cells. Thus the immunized animal's blood may contain, in addition to its naturally occurring antibodies, diverse populations of induced antibodies that are able to bind to different portions of the antigen molecule. Bleeding the animal will supply an 'antiserum' enriched in 'polyclonal' antibodies specific for the artificially introduced antigen.

Monoclonal antibodies are initially produced in the same way, by immunizing an animal (usually a mouse or rat). Once polyclonal antibodies are being produced, the plasma cells of the animal's spleen are fused with those of a plasma cell tumour (myeloma) from the same strain of mouse or rat. The myeloma cells can grow and divide in culture but do not produce antibodies. The fused cells have characteristics of both parent cells: they can be cultured and each cell continues to produce its antibody. By a process of subdivision, cultures cloned from single plasma cells are obtained. The antibody secreted by these cells is of one type only, specific for a single epitope, and is known as a monoclonal antibody [7].

1.3. Antigen–antibody reaction

Among the sub-types of Ig molecule, the most useful for immunohistochemistry are of the IgG and IgM classes. IgG molecules are Y-shaped, while IgM molecules consist of a combination of five IgG-like molecules. The two ends of the short arms of the 'Y' are identical antigen-binding areas, with one IgG molecule being specific for one antigenic group. The two arms together are known as the antigen-binding fraction, F_{ab2}. The long arm is the crystalline or constant fraction (F_c). The attraction of the F_c portions of antibodies to receptors in certain cells is part of the natural immune response that is triggered when the body encounters foreign proteins (antigens). The F_c portion is the same for all IgG molecules of one sub-class from any one animal species and carries an immunogenic sequence on each of its two identical amino-acid chains. When an animal is immunized with IgG from another species, antibodies will be raised preferentially (but not exclusively) to the F_c part of the molecule. This is important for immunhistochemical staining because, to amplify the reaction, the method usually involves more than one layer of antibodies, the antibody of one layer acting as antigen for the next.

1.4. Tissue preparations suitable for immunohistochemistry

Immunohistochemical methods can be carried out on any tissue preparation in which the antigen can be made insoluble while retaining its antigenicity and its tissue context. In practice, this means that the tissue must be fixed (Chapter 6) to preserve both it and the antigen (except in the case of living cell preparations). However, some antigens are very susceptible to fixation, particularly by formalin, which may introduce so many cross-linking (methylene) groups into the antigen molecule that the antigenic sites are no longer available to the antibody. These problems can usually be overcome by use of 'antigen retrieval' methods (see Section 3.4). In general, mild, precipitant fixatives such as alcohol or acetone are useful for preserving large proteins, whereas the cross-linking effects of formaldehyde or glutaraldehyde are needed to make small peptides insoluble.

If the tissue structure is to be well preserved, provided that the antigenicity survives the process, formalin fixation followed by paraffin embedding and sectioning (Chapter 7) is the conventional and best procedure for supplying well resolved antigen localization for the light microscope. However, other types of tissue preparation are also suitable for immunohistochemistry, e.g. free-cell preparations, cultures, fresh or pre-fixed frozen (cryostat) sections (Chapter 6) and (some) resin sections. The main stricture, as for most light microscopical work, is that the preparation must be thin enough to allow the passage of light. Because the penetrative ability of antibody molecules is limited, the best resolution is obtained on thin sections (1–5 μm) or monolayers of cells. Additional thickness does not necessarily mean additional reaction, and if the reaction product is increased too much it may obscure its own specific localization. The thickest preparations that can be dealt with are stretched membranes (e.g. diaphragm or iris) or 40–80 μm thick sections that have been cut with a vibrating or freezing microtome. Special treatment is required, involving long periods of exposure to antibodies and washing buffers. Even for immunofluorescence, with the use of epi-illumination on the fluorescence microscope, thin sections (5 μm) are most suitable. Thicker immunofluorescent preparations are best dealt with by confocal microscopy (Chapters 1 and 2).

1.5. Types of label and their uses

Labels can be attached to antibodies or other proteins by a chemical process (conjugation) whereby the label is attached covalently to the Ig (fluorochromes) or is linked to it by a bridging molecule such as glutaraldehyde (enzyme labels). The ratio of label to Ig is an important consideration. The details are beyond the scope of this chapter but simple methods are described in [8]. It is easy to purchase labelled antibodies, and excellent reagents are available with a wide variety of different labels that will answer most immunohistochemical needs.

1.5.1. Fluorescent labels

Fluorescent labels come in many colours. Among the most popular are FITC with its apple green fluorescence and rhodamine derivatives such as tetramethylrhodamine isothiocyanate and Texas Red, which give red fluorescence. Blue fluorescence can be achieved with 7-amino-4-methylcoumarin acetic acid, and a range of these (and other colours such as pink and yellow) is provided by cyanine dyes, which give a longer-lasting fluorescence and have also been widely used for labelling gene probes. The latest additions to the range of fluorescent labels are the Alexa dyes, supplied by Molecular Probes. Fluorescence is recommended for labelling living cells because it can be incorporated without killing the cells.

A fluorophore emits coloured light after illumination of the preparation with light of a specific excitation wavelength. The fluorescence microscope is provided with filters that allow passage only of light of the required wavelength. In suitable circumstances, two different antigens in a single preparation can be labelled with different fluorochromes and viewed by switching filters to show the localization of first one antigen and then the other.

Immunofluorescence reveals only the labelled structures; the underlying tissue ideally should not emit fluorescence when exposed to the exciting wavelength. Thus it is particularly suitable for cell preparations in which the tissue context is unimportant and for frozen sections in which the tissue structure is probably less well preserved than in paraffin sections. There may be a low level of background fluorescence, which is helpful in allowing the antigen-bearing structures to be localized in the tissue section. Various fluorescent dyes can be used as counterstains for nuclei or other tissue components or , if necessary, to disguise intrusive background fluorescence. The specific immunofluorescence will eventually fade, even in mountants that contain fluorescence preservatives. Formalin-fixed frozen sections whether fixed before or after sectioning, and paraffin sections, commonly have a high level of intrinsic fluorescence and are therefore better suited to immunohistochemistry with enzymic labelling.

1.5.2. Enzyme labels

Enzymes are not microscopically visible as labels for antibodies. After the immunohistochemical reaction there is a further incubation in a suitable substrate and chromogen. The enzyme label catalyses the splitting of the substrate and capture of the chromogen, resulting in the deposition of a coloured, insoluble end-product on the preparation, very near to the site of antibody binding, thus identifying the antigen in the tissue. It is important that the buffer pH and the concentrations of substrate and chromogen are correct to ensure an optimal enzymic reaction.

Because the product is insoluble, permanent preparations can be made, providing an advantage over immunofluorescence. Enzyme-labelled immunohistochemistry is suitable for paraffin or resin sections, as well as for cell preparations or frozen sections. However, one must be aware of the possible presence in the tissue of an enzyme that resembles the one used as a label. Such an endogenous enzyme must, if present, be prevented from reacting with the substrate, to avoid confusion with the applied antibody-conjugated enzyme.

1.5.2.1. Horseradish peroxidase

Horseradish peroxidase was the first enzyme to be used as an antibody label and it has not been superseded. Endogenous peroxidase-like enzymes can be blocked before the immunoreaction, usually by application of excess of the enzyme's substrate, hydrogen peroxide, which irreversibly inhibits the activity of the endogenous enzyme. After the immunoreaction the applied enzyme is 'developed' through its action on a much lower concentration of hydrogen peroxide and a chromogen, 3-3′-diaminobenzidine tetrahydrochloride (DAB) (see Method 7). This is oxidized and polymerizes to give a dark-brown precipitate, which does not dissolve in any solvent or mountant. The deposits can be made black by inclusion of a heavy metal salt, e.g. nickel sulphate, in the incubating medium. The substitution of 3-amino-9-ethylcarbazole for DAB will yield a red product, but this is alcohol-soluble, so a water-based mountant must be used. Counterstaining with alum–haematoxylin provides a contrasting blue nuclear stain.

1.5.2.2. Alkaline phosphatase

This is another popular enzyme label. It has the advantage that pre-blocking of the endogenous enzyme is not required because all alkaline phosphatase iso-enzymes, except the intestinal isoform, which is used as the label, can be blocked by inclusion of an inhibitor, levamisole, in the incubation medium. Alkaline phosphatase is not a suitable marker for intestinal tissue unless the strong staining given by the endogenous enzyme of the brush border of the mucosal epithelium can be ignored.

Alkaline phosphatase splits naphthol phosphate, the naphthol combining with a variety of diazonium salts as chromogens, resulting in bright-blue or red azo dye end-products (alcohol-soluble). Alternatively, a more sensitive reaction with 5-bromo-4-chloro-indolyl phosphate (BCIP) yields a less soluble, purple/brown reaction product (see Methods 8 and 9).

1.5.2.3. Other enzymes

Glucose oxidase, a plant enzyme that is absent from animals, has been used as a label. In animal tissues no pre-blocking is required and a strong, dark-blue coloured product results.

β-D-Galactosidase derived from bacteria is also used as a label; its pH optimum is different from that of the animal enzyme. Therefore, in the absence of bacteria in the preparation, it provides a useful label that can be developed to give several colours, usually a permanent turquoise–blue.

1.5.3. Colloidal gold

Antibodies are adsorbed to colloidal gold particles, instead of being chemically conjugated to them. This label was originally used (and still is) for ultrastructural immunocytochemistry because of its particulate form and high electron density. As a light microscopical label its intrinsic magenta colour can only be seen if highly concentrated antibody is used. However, lower concentrations of colloidal gold may be made visible by treatment with a suitably compounded solution containing a silver salt, which precipitates intensely black metallic silver onto the gold particles [9,10]. This is an extremely sensitive immunoreaction.

1.5.4. Biotin

This small molecule has become an important label because of the increased sensitivity it imparts to immunohistochemical methods by its strong combination with the much larger glycoprotein, avidin. Neither avidin nor biotin are visible microscopically, but each can be labelled with enzyme, colloidal gold or any other label. One avidin molecule is able to carry a large number of smaller label molecules. Several biotin molecules can be attached to an Ig molecule without significant impairment of its immunoreactivity. Usually this biotin is not labelled but binds a further layer of labelled avidin or a combination of unlabelled avidin with labelled biotin [avidin–biotin complex (ABC)].

1.6. Background problems and their elimination

1.6.1. Endogenous enzymes

Frozen sections or cell preparations, and even fixed paraffin sections, retain enough peroxidase-like or alkaline phosphatase activity to be intrusive in an immunoperoxidase reaction. Blocking methods are described in Sections 3.3.2 (peroxidase) and Sections 1.5.2.2. and 4.4 (alkaline phosphatase).

1.6.2. Non-specific staining

A frequent problem in immunostaining is a high level of 'background' coloration owing to unwanted binding of antibodies to tissues in addition to the specific immunoreaction. This is usually classed as non-specific, because of the attraction of Igs to F_c receptors (fresh tissues only; the receptors are destroyed by formalin fixation) or to charged or hydrophobic sites. Occasionally, there are specific but unwanted immunoreactions between antibodies in the primary antiserum and antigens in the tissue. Non-specific binding is usually eliminated by saturating the tissue preparation with normal (non-immune) serum before applying the primary antibody. Igs in the normal serum will occupy F_c receptors and other non-specific binding sites and prevent binding of applied antibodies from the first and subsequent layers. It is important to use the serum from the species that provides the second antibody in any two- or three-layer technique, e.g. normal goat serum if the second antibody is goat anti-rabbit Ig or goat anti-mouse Ig (see Section 2). In a one-step, direct method, use serum from the same species as the labelled antibody. If the tissue preparation is processed to paraffin sections (no active F_c receptors), bovine serum albumin, or another protein such as casein that does not take part in the reaction, can be substituted for normal serum. The blocking protein is applied before the primary antibody.

1.6.3. Biotin blocking

When avidin–biotin methods are used, an additional blocking procedure may be required for endogenous biotin, which would bind to labelled avidin and be revealed in the final development. Biotin is particularly prominent in the liver and kidney, and its reactivity may become enhanced in these, and in other tissues, after antigen retrieval (Section 3.4). Biotin blocking consists of applying avidin to bind to any biotin in the tissue, and then adding biotin to block any free biotin-binding sites on the avidin, or any endogenous avidin. Because biotin has only one binding site for avidin, no further avidin- or biotin-labelled reagents can bind [11]. Blocking can be carried out at any stage prior to application of a biotin-labelled reagent, but it is usually done after the normal serum application. If biotin-labelled probes are being used in an *in situ* hybridization method, endogenous biotin must be blocked at some stage before the probe is applied. A biotinylated probe can be visualized with labelled avidin, or with a labelled antibody to biotin, or an unlabelled one, which is subsequently identified with one of the methods detailed in Section 2.3.2.

Method 1. Biotin blocking

This method is incorporated into the immunostaining method described in Method 6.

Reagents

- 1 mg/ml avidin in phosphate-buffered saline (PBS) or Tris-buffered saline (TBS); see section 3.2
- 0.1 mg/ml biotin in PBS or TBS

Procedure

1. Apply avidin and incubate for 15 min at room temperature.
2. Rinse in PBS.
3. Apply biotin and incubate for 15 min at room temperature.
4. Rinse in PBS.

1.7. Optimal antibody concentrations

It is imperative to establish the 'correct' antibody concentration for your particular needs. This is the dilution for each antibody layer that will saturate the antigen available. A monoclonal antibody is usually suitable for immunocytochemistry at a concentration of 1–10 μg/ml. The dilution of a polyclonal antiserum or purified Ig will probably be in the range 1:50 to 1:8000. The supplier's data sheet may give a guide to a suitable dilution, but you will have to test the antibody (using the immunostaining method you intend to use for the same standard incubation times) in a series of (doubling) dilutions against a known positive control that contains a sufficient amount of the antigen under consideration and, most importantly, that is prepared in the same way as the test tissue. Polyclonal antibodies often give the best results at a high dilution to remove minor populations of unwanted antibodies. An overnight incubation (at 4 °C to prevent evaporation) will then allow both the high- and low-avidity specific antibodies in the antiserum to attach to the antigen. Choose the dilution that provides the most intense positive stain, combined with a low non-specific background. In this way you will have selected a concentration that will saturate a large amount of antigen and that will, therefore, be adequate to demonstrate the antigen in your test preparations, whether it is present in small or large amounts.

Monoclonal antibodies do not require high dilution because there is no extra population of unwanted antibodies, and binding can be sufficient in a shorter time (e.g. 1 h at room temperature). Some monoclonal antibodies bind slowly and require a longer incubation period. However, economy dictates that they should be used at as high a dilution as is compatible with good immunostaining.

Labelled second antibodies should be tested against a known first antibody at its established optimal dilution. Commercially labelled antibodies are usually purified. Dilution is therefore less important for removal of unwanted antibody populations, but it is still required to reduce non-specific binding and for economy. Establish a dilution that will give results in 30–60 min. Once an optimal dilution has been established, it should suffice for use with any primary antibody of the correct species.

2. Methods

In this description of the steps of the various methods it is assumed that all preliminary blocking steps have been carried out. The label can be fluorescent, an enzyme or gold. Practical aspects are described in Section 3.

2.1. Direct method (one step)

Labelled antibody is applied directly to the tissue preparation and allowed to react. Unbound antibody is rinsed off and after development (using either enzyme or gold labels) the preparation is viewed in a suitable microscope. The ratio of bound antibody molecules to antigen molecules is approx. 1:1 (Fig. 1).

2.2. Indirect method (two steps)

The primary antibody bound to the tissue antigen is unlabelled and acts as an antigen for the second layer. This second layer is a labelled antibody from another animal species that has been raised to Ig from the species that provided the first antibody. The indirect method has two or more times the sensitivity of the direct method because an Ig molecule acting as an antigen has at least two identical binding sites for antibody on the F_c fragment (Fig. 2).

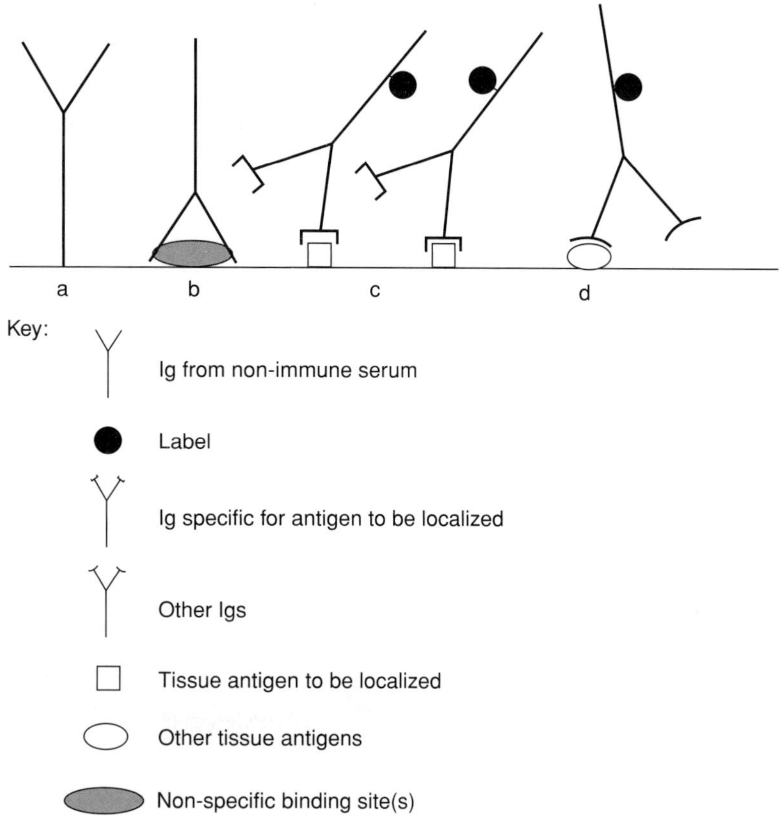

Fig. 1. The direct method

(a) F_c receptors are occupied by Ig from non-immune serum; (b) non-specific binding sites are occupied by Ig from non-immune serum; (c) labelled antibody binds specifically to its tissue antigen; (d) other specific labelled antibodies in polyclonal antisera can bind to tissue antigens but are usually present in low concentrations. This is not a problem with monoclonal antibodies.

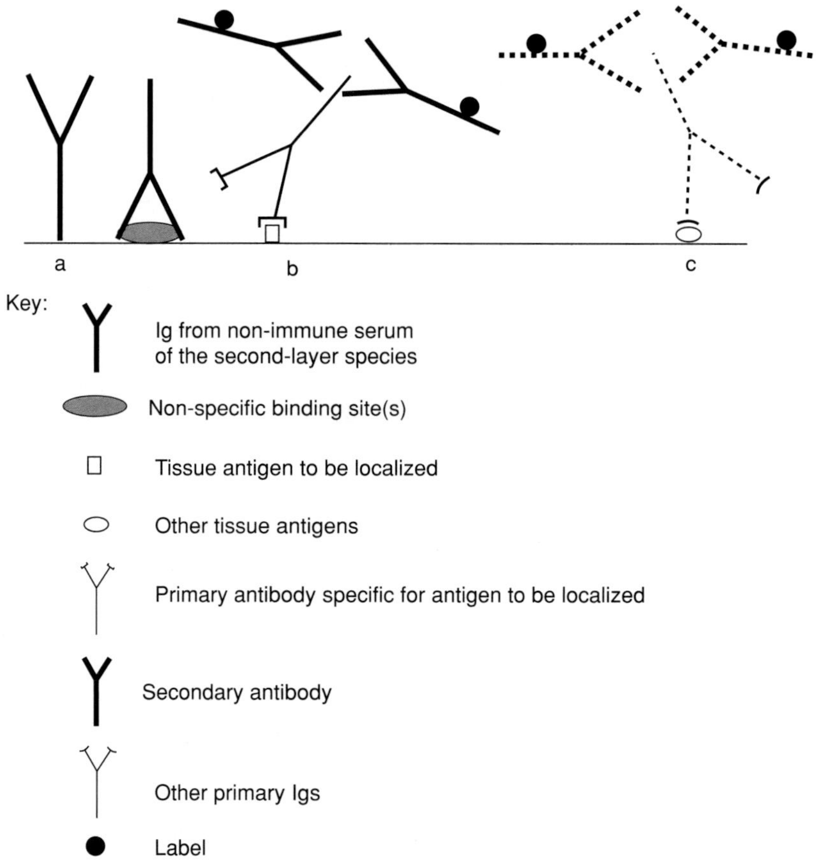

Fig. 2. The indirect method

(a) F_c receptors and non-specific binding sites are occupied by Ig from non-immune serum from the second-layer species (e.g. goat); (b) unlabelled antibody from the first species (e.g. rabbit) binds specifically to its tissue antigen and is localized with the labelled antibody of the second layer (e.g. goat anti-rabbit Ig); (c) other specific antibodies in polyclonal antisera can bind to tissue antigen and can be localized by labelled second antibody, but they are more dilute than in the one-layer method. This is not a problem with monoclonal antibodies.

2.3 Three-step methods

These provide even greater sensitivity, allowing a small quantity of antigen to be visualized more easily than with the one- or two-step methods. Even if the increased sensitivity is not required to visualize the attachment site of the primary antibody, the three-layer techniques allow greater dilution of the primary antibody and, therefore, removal of unwanted and non-specifically bound components. This is particularly important when polyclonal antisera are being used. Three-step methods also allow the most economical use of the primary antibody, which is likely to be the most expensive reagent applied to the preparation.

2.3.1. Unlabelled antibody–enzyme complex [peroxidase–anti-peroxidase (PAP)] method

Both the primary and secondary antibodies are unlabelled, therefore their immunoreactivity is unhindered by label molecules. The second antibody is applied in excess over the primary so that the maximum number of secondary antibody molecules are bound, each having an antigen-binding arm free for the third-layer reactant. The third layer is an Ig from the same species as the primary antibody, which has been raised against horseradish peroxidase and then combined immunologically with this enzyme. Two Ig molecules now share three peroxidase molecules and are held together in a stable, cyclic structure — the PAP complex [12]. This is possible because peroxidase has several binding sites for antibodies whereas an antibody has only two binding sites for its antigen (Fig. 3). Thus the molecular ratio of peroxidase label to tissue antigen should be at least 3:1. The enzyme itself has full activity because, being bound immunologically, it has not been altered by chemical conjugation.

2.3.1.1. Other enzyme–antibody complexes

Complexes similar to PAP can be made with alkaline phosphatase [alkaline phosphatase–anti-alkaline phosphatase (APAAP)] or glucose oxidase.

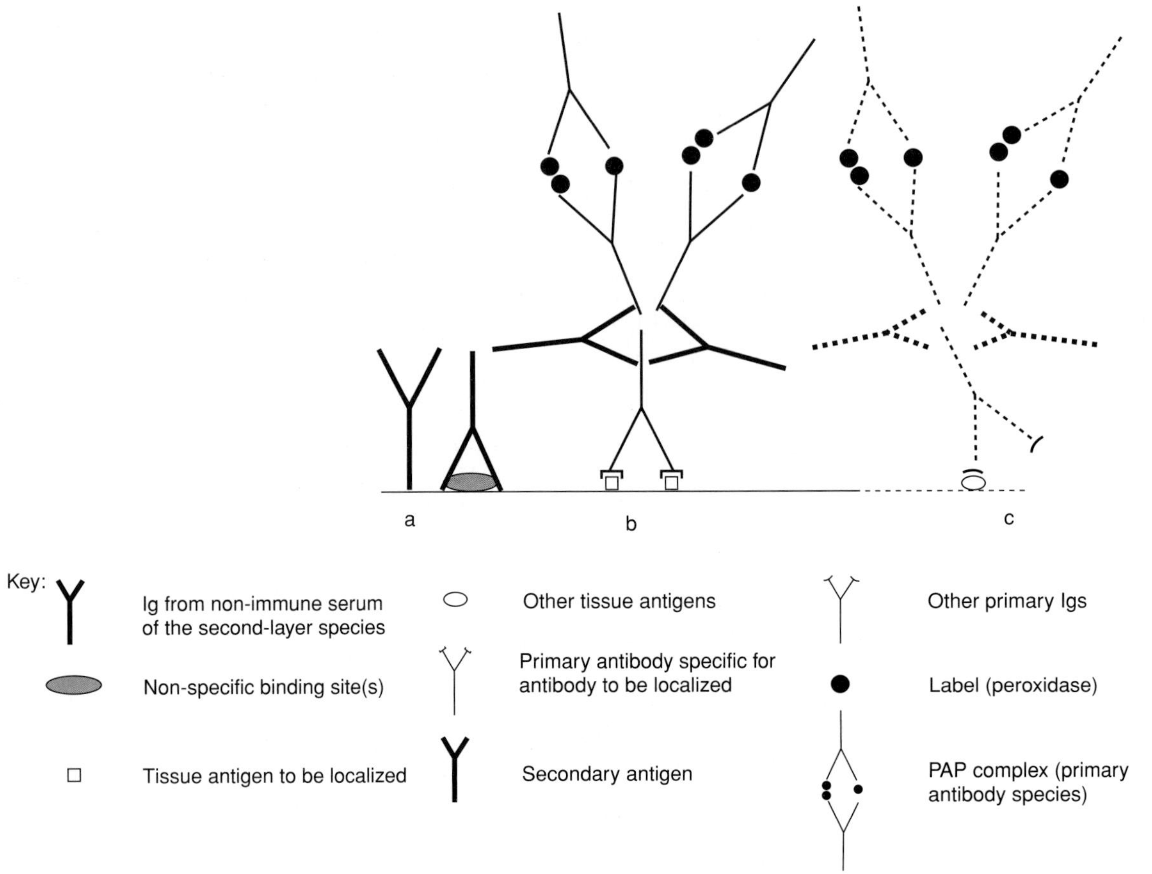

Fig. 3. The PAP method

(a) F_c receptors and non-specific binding sites are occupied by Ig from non-immune serum from the second layer species (e.g. goat). (b) Unlabelled antibody from the first species (e.g. rabbit) binds specifically to its tissue antigen and then acts as antigen for the second layer of unlabelled antibody (e.g. goat anti-rabbit Ig), which is applied in excess. This antibody, with its free F_{ab}, binds the third layer, the rabbit Ig–peroxidase complex (PAP). (c) Other specific antibodies in polyclonal antisera can bind to tissue antigen, but they are much more dilute than in the one- and two-step methods. This is not a problem with monoclonal primary antibodies.

2.3.2. Avidin–biotin methods

Avidin is a large glycoprotein that is usually extracted from the white of eggs. It has four binding sites for biotin. Avidin has certain disadvantages: it may attach to lectins in tissues through its carbohydrate groups, and its strongly basic isoelectric point ($pK_i = 10$) may cause non-specific binding to negatively charged sites in the tissue. Most manufacturers now supply a modified form of avidin in which these problems have been overcome. Alternatively, avidin is replaced by streptavidin, a bacterial protein with similar properties but with uncharged molecules and no carbohydrate groups.

Biotin, a very small molecule, is a vitamin present in quantity in the yolk of eggs. A biotin molecule has one binding site for avidin and other sites for chemical conjugation to labels such as peroxidase.

2.3.2.1. Labelled avidin method

The avidin itself may be labelled and applied to the immunoreaction as a third layer after a biotin-labelled secondary antibody (Fig. 4).

2.3.2.2. ABC method

Prior to use in the immunoreaction, labelled biotin is combined with avidin to form a complex in which many labelled biotin molecules are held together by several molecules of avidin. The quantities of avidin and labelled biotin to be combined are calculated by the supplier so that some unoccupied biotin-binding sites remain on the avidin of the complex, for binding to the biotin on the antibody (Fig. 4).

The high sensitivity and versatility of reactions involving avidin (or streptavidin) and biotin are due to the large number of biotin molecules that can be conjugated to antibodies, the large number of label molecules that can be conjugated to avidin and the almost irreversible binding between avidin and biotin (affinity constant 10^{-15} mol^{-1}). Thus, the ABC reaction, with a choice of visible labels, has become one of the most popular immunohistochemical methods [13].

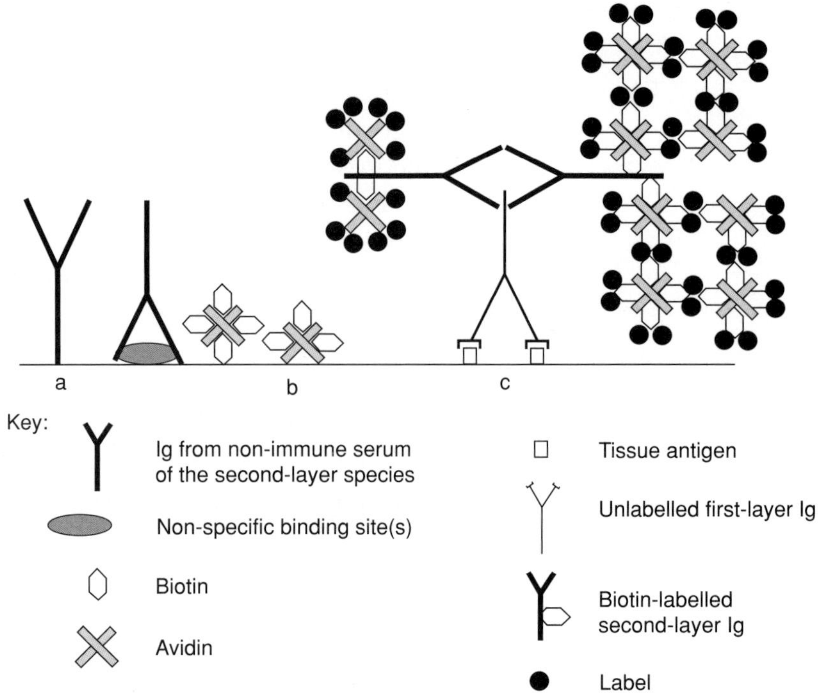

Fig. 4. Labelled avidin and ABC methods

(a) F_c receptors and non-specific binding sites are occupied by Ig of non-immune serum from the second-layer species (e.g. goat). (b) Endogenous biotin and avidin are blocked by the application of avidin followed by biotin. (c) Unlabelled antibody from first species (e.g. rabbit) binds specifically to its tissue antigen and is localized with the second-layer biotin-labelled antibody (e.g. goat anti-rabbit Ig). The third layer comprises labelled avidin (shown on the left) or a complex of avidin combined with a biotinylated label (ABC). This method is so sensitive that the primary antibody can be extremely dilute, thereby reducing the problem of minor populations of unwanted antibodies.

3. Practical aspects

In this section it is assumed that the preparations to be immunostained are mounted on glass microscope slides. Adaptations may be necessary for cell preparations grown on coverslips or in wells.

3.1. Equipment

- Humid chamber
  - This can be purpose-made or as simple as a large, flat-bottomed, plastic Petri dish (e.g. Sterilin, 14 cm diameter) containing a small wad of damp paper tissue and a rack made of two wooden applicator sticks or glass rods, to support the slides off the bottom of the dish and avoid the flow of liquid under and around the slides.
- Pipettes
  - Gilson or other adjustable volume pipettes, or Pasteur pipettes.
- Hydrophobic pen (optional)
  - This is useful for isolating the section or cell preparation on the slide to make application of antibodies easier. It is possible to do this without a hydrophobic pen by wiping the slide dry around the tissue preparation to create a difference in surface tension between the damp tissue area and the dry slide. However, there is a greater danger of the preparations drying during the procedure or being wiped off with this method.
- Paper tissue (Kleenex or similar)
  - This is to be used for drying round the preparations and removing excess buffer from them.
- Small containers
  - For diluting and storing antibody solutions. Eppendorf or similar conical plastic microcentrifuge tubes (Fisher Scientific), or autoanalyser vials and caps (LIP Ltd, Catalogue numbers 26104 and 26146) are suitable. Wrapping in paraffin or polyvinyl film will provide additional protection against evaporative losses during storage.

- Coplin jars
  - These have ridges to hold slides. Alternatively, place slides in carriers in containers of buffer for washing the preparations.

3.2. Buffers, antibody diluent and storage

3.2.1. Buffers

Before the reaction is started and between the additions of new layers, the preparations will have to be rinsed in isotonic buffer. PBS) or Tris-buffered saline (TBS) are most commonly used. PBS is less expensive to make up than TBS (not if tablets are used). TBS is said to prevent some background staining, owing to its high ionic content, but this is not usually a problem. TBS is preferred for alkaline phosphatase-labelled reactions. Where PBS is specified in this chapter, TBS may be substituted.

- PBS (pH 7.0–7.4)
  - 1000 ml of water
  - 0.15 M NaCl; 8.7 g
  - 2 mM KH_2PO_4; 0.272 g
  - 8 mM $Na_2HPO_4 \cdot 2H_2O$; 1.41 g

  Dissolve salts separately, with heat if necessary, then mix and check the pH.

  PBS can be made up as a concentrated (10×) stock solution to be diluted for use. Check the pH of the diluted solution, not the concentrate, because the pH changes on dilution.

- TBS (pH 7.6)
  - 0.15 M NaCl; 8.7 g
  - 0.05 M Tris(hydroxymethyl)methylamine; 6.02 g
  - 3.0–3.2 ml of HCl (concentrated)

  Dissolve Tris and NaCl in 950 ml of distilled water. Adjust the pH to 7.6 with concentrated HCl. Adjust the volume to 1 litre with distilled water.

3.2.2. Antibody diluent

Dilute antibodies that are not enzyme-labelled in a solution of PBS containing 0.1% sodium azide as a preservative and 0.1% bovine serum albumin (BSA).

The BSA acts as an 'inert' protein in relatively high concentration that competes with the antibody molecules for 'sticky' binding sites on the walls of the storage vessel. It does not interfere with specific immunoreactions. Antibodies can be diluted in simple PBS (or TBS) for immediate use.

Note *Do not dilute enzyme-labelled reagents in this diluent because azide is an enzyme inhibitor. Use PBS alone (not for long-term storage) or a proprietary enzyme preservative such as Protexidase (ICN Biomedicals Ltd) to store your enzyme conjugates ready for use. Do not store dilute solutions of fluorescent conjugates; there is some danger of the label becoming detached from the antibody.*

3.2.3. Antibody storage

Follow the supplier's recommendation for storage. Many antibodies can be stored at their supplied concentration at 4 °C. Some, particularly monoclonal antibodies, may survive better if snap-frozen in aliquots suitable for diluting to the working concentration, and stored at or below −20 °C (see Section 3.2.3.2).

3.2.3.1. Antibodies stored at their working dilution

It is convenient to store antibodies at their working dilution at 4 °C, provided that they will remain active under these conditions. They should be diluted in antibody diluent (see Section 3.2.2).

3.2.3.2. Antibodies stored frozen

Store polyclonal antibodies in frozen undiluted aliquots, or, at a 1:10 to 1:100 concentration in antibody diluent (with BSA; see Section 3.2.2). Freeze monoclonal antibodies in undiluted (as received from supplier) aliquots in diluent or in culture medium containing protecting proteins, such as those in foetal calf serum.

> ***Note*** *Do not freeze antibodies slowly by placing aliquots directly in the freezer. Snap-freeze them in liquid nitrogen then store at −20 °C or a lower temperature. Do not thaw and re-freeze them repeatedly.*

3.2.3.3. Storage at low temperature without freezing

Dilute the antibody 1:1 (v/v) with glycerol and place at −20 °C. The solution will not freeze. Remember that it is already diluted when you make your working solution.

3.3. Preliminary preparation of tissues

3.3.1. Fixation

These recommendations are very general because the antigen to be localized may determine the fixation to be used.

3.3.1.1. Fresh tissue preparations

Cells

Once cell preparations are made, fix them immediately in a mild, protein-precipitating fixative such as alcohol or a 1:1 (v/v) mixture of acetone and methanol for a few minutes. Experiment with keeping the preparations in alcohol until they are used or placing them in PBS for storage. They may also be air-dried after fixation if this proves suitable.

Some antigens may require formaldehyde fixation, e.g. in 4% paraformaldehyde (pH 7.4) for 30 min.

Frozen sections

Allow fresh frozen cryostat sections to dry at room temperature for at least 1 h (overnight is often better) so that they adhere well to the slide and partly denature. Then fix for 10 min in acetone. Allow the sections to air-dry before staining.

Some antigens may require formalin fixation. A period of 10–30 min at room temperature is usually adequate. After this, do not allow the cells or sections to dry but store them in PBS until they are to be stained.

> **Note** *Acetone should be stored over molecular sieves to keep it water-free and re-usable several times.*

Storage of fresh tissue preparations

You can store these at −20 °C (or lower), air-dried and unfixed, or after acetone or alcohol fixation. Wrap the slides in cling-film or aluminium foil (back-to-back to avoid damaging the sections), seal them in a plastic bag with some silica gel or other desiccant and place them in the freezer. Allow the entire sealed packet to come to room temperature before opening so that the desiccant absorbs any condensed water. This is particularly important for unfixed material because water condensing on the preparations could cause antigens to dissolve and diffuse.

3.3.1.2. Fixed tissue

Tissue fixed in formalin or another fixative is processed by a standard method through graded alcohols to an intermediate solvent and then impregnated with paraffin wax and embedded in a wax block (Chapter 7). Prolonged formalin fixation is not advisable for immunocytochemistry. It may result in so much internal protein cross-linking that epitopes are altered and are no longer available to the antibody. Various antigen retrieval techniques can be applied to regenerate them (see Section 3.4), but it is better to avoid this option. For a block measuring 0.5 cm × 0.5 cm × 0.5 cm, fixation for 12–24 h is sufficient.

3.3.1.3. Sectioning

Immunostaining is best seen in thin preparations because the reactions take place on the surface of the section. Sections 2–5 μm thick are suitable. Mount the sections on clean glass microscope slides that have been treated to optimize tissue adhesion (see Method 2).

Paraffin sections

After picking up paraffin sections, allow them to dry overnight at 37 °C to ensure good adhesion. To save time, the mounted sections can be heated for 10–20 min in an incubator at 60 °C to just melt the wax without damaging most antigens. Do not overheat them on a hot plate. Cool the slides before de-waxing and bringing to water through graded alcohols in the usual way (Chapter 7).

Frozen sections

See Section 3.3.1.1.

Method 2. Slide coating

Mount frozen sections, cell preparations and paraffin sections that do not require heat-mediated antigen retrieval on poly-L-lysine-coated slides. This gives a positive charge to the glass, which attracts negatively charged tissue components. Paraffin sections requiring heat-mediated antigen retrieval should be mounted on silanized slides or proprietary slides that give stronger adhesion, e.g. ‘polysine’ (Merck), Vectabond (Vector Laboratories) or similar slides.

Poly-L-lysine coating

Materials

- Poly-L-lysine hydrobromide (MW 150000–300000 kDa) (Sigma P1399)
- Distilled water

Procedure

1. Make up a 0.1% solution of poly-L-lysine hydrobromide in distilled water. Place 1 ml aliquots in a freezer for storage. Allow the solution to thaw completely and mix well before use. Unused solution can be re-frozen. Ready-to use solutions (Sigma P8920) and pre-coated slides are also available.

2. Place a small drop (5–10 μl) at the end of a slide. Rock the short edge of another slide in the drop to spread it along the width of the slide, then push the fluid along the length of the slide to create a thin layer of poly-L-lysine over the entire surface.

   Interference colours should be visible as the solution is spread. If the solution forms droplets as it dries, discard the slide and ensure that the rest of the slides are clean. If necessary, wash them in acid–alcohol (1.0% HCl in 70% ethanol), rinse in alcohol and allow to dry before coating.

3. Mark the coated side. Coated slides can be stored indefinitely at room temperature.

Silane coating

Materials

- Aminopropyltriethoxysilane (APES) (Sigma A3648)
- Acetone

Procedure

1. Soak slides in acetone for 5 min.

2. Immerse the slides in 2% APES in acetone for 5 min. The APES solution in acetone keeps for only 1 day.

3. Rinse in two changes of acetone for 5 min each and allow to dry. Coated slides can be stored indefinitely at room temperature.

3.3.2. Blocking endogenous peroxidase-like enzymes

Blocking is a necessary procedure if antigens are to be identified with peroxidase and if there is significant peroxidase-like activity in the preparation. This can be checked by placing a spare preparationn in the peroxidase developing solution (Method 7). If there is no intrinsic activity, there is no need to carry out the blocking step, which can damage some antigens, particularly in fresh cell preparations or frozen sections.

Peroxidase blocking can be performed before or after antigen retrieval (see below). If it damages an antigen, it can be done after incubation with the primary antibody, at any stage before the peroxidase-labelled reagent is applied.

3.3.2.1 Paraffin sections

After bringing sections to water, immerse the slides in 0.6% (w/v) H_2O_2 (in tap water, distilled water, PBS, TBS or methanol) for 15 min. A 2% (v/v) solution in tap water, buffer, distilled water or methanol can also be used. Then rinse the slides in water and leave them until ready to continue.

3.3.2.2. Fresh cell preparations or frozen sections

These often require more gentle treatment. A 0.3% (w/v) H_2O_2 solution, diluted in methanol (or in 70% methanol in PBS) should be used. Apply this for 30 min, then transfer to PBS.

A longer method, but better for preserving some cell surface antigens, uses nascent H_2O_2, produced by the action of glucose oxidase on glucose, combined with the inhibitory effect of azide [14]. The solution consists of PBS at 37 °C containing 1 mM sodium azide, 10 mM glucose and 1 unit/ml glucose oxidase. The solution should be prepared just before use. Immerse the preparations for 1 h at 37 °C, then rinse in PBS and continue with the procedure.

3.4. Antigen retrieval

The fixative action of formaldehyde is produced by the formation of cross-linking methylene bridges within protein molecules. These may alter the conformation of some antigen molecules so that antibodies do not recognize them. Sometimes epitopes can be made available again in the preparation by subjecting it to some kind of 'antigen retrieval', which is thought to break these bonds. The most usual retrieval methods are protease treatment (usually with trypsin or protease XXIV) or the increasingly popular method of super-heating sections in citrate or other buffer in a microwave oven or pressure cooker [15–17]. Surprisingly, this does not seem to damage antigens or tissue structure, provided that the tissue is properly fixed. The buffers used are all capable of chelating calcium and are thought to remove tissue calcium from its bonds with formalin after it has been partially released by the heating part of the procedure. Because the calcium is permanently removed the bonds do not re-form and the antigenic sites remain revealed [18]. The pH of the retrieval solution may also be influential [19,20]. However, it appears that heat is the most important factor in antigen retrieval, and the higher the temperature, the shorter the heating time required. If sections tend to become detached from the slides, they may be heated in buffer for several hours (or overnight) at a lower temperature, e.g. 60–80 °C. Shi et al. [21] provide a full review of antigen retrieval for immunohistochemistry. Fresh preparations and tissues fixed in precipitant fixatives do not usually need antigen retrieval.

There is no strict rule as to the type of antigen retrieval suitable for different antigens. If poor results are obtained without pretreatment, the literature on immunocytochemistry for similar antigens should be consulted and the standard methods attempted. In general, the longer the tissue has been in fixative, the longer will be the time required for antigen retrieval. In the methods described below, standard times are used. These are 10 min for trypsin pretreatment (this can be extended to 30 min or longer), 1–10 min for protease XXIV pretreatment, 10 min for microwaving in citrate buffer (this can be extended to 20 or 30 min) or 2 min in a pressure cooker followed by cooling for 10 min. The time of incubation in enzymes is more critical than heating time, as prolonged exposure may damage the tissue. Each laboratory will have its own methods and it is important to standardize the methods for your laboratory so that you always start from the same temperature and volume and, if necessary, make adjustments from that standard. In the methods that follow, from the author's laboratory, it is assumed that the paraffin sections have been brought to water.

Method 3. Antigen retrieval by trypsin pretreatment

Trypsin is a relatively cheap enzyme, so it is convenient to place the slides in a rack and immerse them together in a sufficient volume of freshly prepared solution.

Materials

- Trypsin
  - Any crude porcine trypsin containing some chymotrypsin, which is probably the most active ingredient. DIFCO 125 is satisfactory.
- Calcium chloride
  - 0.1% in 0.005–0.01 M Tris/HCl buffer (pH 7.6–8.0), kept as a stock solution at 37 °C.
- n 1.0 M Sodium hydroxide

Procedure

1. Make a fresh solution of 0.1% trypsin in the warm, buffered calcium chloride solution and immediately adjust the pH to 7.8 with sodium hydroxide.
2. Immerse the slides and incubate at 37 °C for the required time (usually 10–20 min).
3. Stop the reaction by rinsing the preparations in tap water.

Method 4. Antigen retrieval by protease XXIV treatment

As protease XXIV is expensive, incubation should be carried out on the slide by covering the section with a drop of the solution.

Materials

- Bacterial protease XXIV (Sigma P8038)
- PBS, kept at 37 °C

Procedure

1. Make a fresh solution of 0.5% protease in warm PBS. It is not necessary to adjust the pH.
2. Apply to the sections and incubate in a humid chamber (see Section 3.1) at 37 °C for the required time (usually 1–10 min).
3. Rinse in tap water.

Method 5. Heat-mediated antigen retrieval in a microwave oven

Materials

- Microwave oven (domestic, with revolving plate, timer and 750–800 W supply)
- 0.01 M Citrate buffer
  - 0.01 M Sodium citrate brought to pH 6.0 with 2 M sodium hydroxide. This buffer will keep well at room temperature for several days.
- Plastic container with lid (e.g. sandwich box)
- Plastic slide rack

Procedure

1. Heat a standard volume (e.g. 500 ml) of buffer in the plastic container and a standard volume of distilled water (e.g. 200 ml) in a plastic beaker, for a standard time (e.g. 2 min at 750 W) in the oven. This ensures that you

always start your microwaving at the same temperature and have a supply of warm distilled water for replacing water lost from the buffer by evaporation. The volume of buffer should be sufficient to cover the slides in the rack.

2. Remove the beaker of water from the oven.

3. Immerse the rack of slides in the buffer and cover the container loosely with its lid. Centre the box on the revolving plate.

4. Heat in the microwave for 5 min. The solution should be boiling after approx. 1 min. Check the buffer level and add warm distilled water to the 500 ml mark if necessary.

5. Repeat for a further 5 min.

   If this process is to continue, top up to the initial volume with water again and repeat for further 5 min periods as required.

6. At the end of the microwaving period, remove the container and cool the preparations by gently running in cold tap water.

   Caution *Care should be taken when removing the container as it is very hot.*

7. Rinse in tap water.

   Note *In some laboratories, preparations are left to cool in the buffer, but we have found that cooling with water is satisfactory and quicker. Never remove sections from the hot solution — the section will dry, buffer will precipitate and artefacts will result.*

4. Immunostaining

4.1. Antibody application

After pre-treatment comprising endogenous peroxidase blocking and antigen retrieval (if necessary), place preparations in PBS.

> ***Note*** *It is important not to let preparations, particularly frozen sections and cell preparations, dry when they have become wet with an aqueous solution; paraffin sections are more robust. Therefore, handle slides one at a time until you become proficient.*

Method 6. Antibody application

Materials

- Humid chamber (Section 3.1)
- Blocking serum (Section 1.6.3)
- Antibody diluent (Section 3.2.2)
- PBS
- Antibodies

Procedure

1. Shake off excess buffer and wipe the slide dry around the preparation. You may wish to isolate it by drawing round it with a hydrophobic pen now if this has not been done already. Place the slide in a humid chamber.
2. Apply a drop of background blocking serum from the same species that provides the secondary antibody (or the primary antibody in a one-layer, direct method) diluted 1:20 in the antibody diluent. Cover the dish and leave for 10 min (or longer as there is no time limit) at room temperature.

3. Do not rinse off the serum, simply drain it off with paper tissue. Ensure that the section remains damp with serum and that the area of slide around it is dry.

> ***Note*** *Blocking of endogenous biotin can be performed at this stage or at any stage before application of the biotinylated Ig (see Section 1.6.3 and Method 1). After blocking, rinse briefly in PBS.*

4. Apply a drop of the primary antibody at its optimal dilution in antibody diluent (see Section 1.7 for details on optimizing antibody dilutions). Cover the dish and leave for the standard time at room temperature or 4 °C (see Section 1.7).

5. Drain off the primary antibody and immerse the slide in PBS. Rinse in three changes of PBS for 5 min each.

6a. *Direct method*. Proceed to Step 10.

6b. *Two- and three-layer methods*. Dry around the section as described in Step 1, removing excess buffer but ensuring that the section remains damp. Replace it in the humid chamber and apply a drop of secondary antibody at its optimal dilution. Cover the dish and leave for the standard time (usually 30–60 min) at room temperature.

7. Rinse in PBS as in Step 5.

8a. *Indirect two-layer method*. Proceed to Step 10.

8b. *Three-layer methods*. Dry around the section as described in Step 1 and apply third layer reagent. Incubate at room temperature for the standard time (30–60 min).

9. Rinse three times in PBS as in Step 5.

10a. *Enzyme preparations*. Develop the enzyme label (Methods 7–9) and complete the finishing procedure.

10b. *Fluorescence preparations*. Mount in a suitable aqueous mountant (Section 4.2).

4.2. Mounting immunofluorescence preparations

Mount preparations in a fluorescence-free aqueous mountant, e.g. glycerol/PBS (9:1, v/v). A good commercially available one that dries hard in 15 min and contains a fluorescence preservative is Permafluor (Thermo Shandon, 434980) and there are many others. View in a fluorescence microscope at a suitable excitation wavelength (495 nm for fluorescein, 530 nm for rhodamine derivatives).

4.3. Peroxidase development

The method given here uses DAB as the chromogen to give an insoluble, dark brown end-product, but there are many other methods.

Note *DAB is a potential carcinogen. Wear gloves, do the reaction in a designated fume cupboard to avoid aerosol inhalation and dispose of the reagent safely by oxidizing it with a few drops of household bleach (sodium hypochlorite). When the solution has become black, pour it down the drain in the fume cupboard with a copious amount of water. Be careful to wash the DAB container thoroughly to remove all traces of bleach which might contaminate the next DAB solution in the container. Alternatively, oxidize the DAB by adding it to an excess of an acidified permanganate solution. While stirring add 15 ml of concentrated sulphuric acid slowly to 85 ml of water, then add 4 g of potassium permanganate. Pour in the used DAB solution and leave overnight. Neutralize with sodium hydroxide before discarding The sodium hydroxide should be added carefully as the temperature will rise.*

Method 7. Colour development from peroxidase-labelled antibodies

Materials

- PBS
- DAB (e.g. Sigma D5637)
- 30% (w/v) Hydrogen peroxide

Procedure

1. Dissolve 50 mg of DAB in 100 ml of PBS — the solution should be a light straw colour.
2. Just before use add 33 μl of 30% hydrogen peroxide (final concentration 0.01%) and mix well.
3. Immerse the preparations for 10 min.
4. Transfer them to 10 ml of PBS and check the reaction microscopically. If necessary, return them to the DAB for a few minutes.
5. Rinse in tap water.
6. Counterstain with alum–haematoxylin (Chapter 8) to give a light blue nuclear stain to contrast with the brown immunoreaction product, differentiate in acid–alcohol, place in tap water or a mildly alkaline solution until the nuclear stain is blue (see Chapter 8), dehydrate, clear and mount in synthetic mountant.

Notes

(a) The developing solution can be used several times and will keep for between 1 and 2 h at room temperature (preferably in the dark). If it becomes brown, discard it.

(b) The final concentration of the substrate, hydrogen peroxide, should be 0.01–0.03% (w/v). If your stock solution is less than 30%, adjust the added volume accordingly. The stock solution of hydrogen peroxide should be stored at 4 °C.

(c) Many companies provide kits for the development of DAB or other chromogens. These are convenient and reliable if you are dealing with a few slides at time, but always more expensive than making your own solutions. Furthermore, they are supplied in small volumes, requiring the DAB to be applied to sections individually rather than immersing the slides. This means that timing the reaction for a large number of slides is difficult and there is more danger of contaminating equipment with DAB. A safe and convenient way of ensuring consistent DAB solutions and avoiding the possible danger of weighing small quantities of DAB is to buy it in pre-weighed amounts, to dissolve it at a concentration of 50 mg/ml in distilled water and to store it frozen in aliquots of 1–5 ml. A volume of 1 ml will give 100 ml of developing solution. Excess DAB stock solution from an aliquot can be refrozen [22].

4.4. Alkaline phosphatase development

Method 8. Development of a red reaction product from alkaline phosphatase

This method uses naphthol AS-MX phosphate as substrate and Fast Red TR salt as the chromogen. Other naphthol phosphates and diazonium salts provide differently coloured end-products.

Many companies provide convenient kits that use various methods for alkaline phosphatase development.

Materials

- Substrate stock solution
  - 0.1 M Tris/HCl (pH 8.2)
  - Naphthol AS-MX phosphate, sodium salt (Sigma N5000)
  - *N*,*N*-Dimethylformamide

Dissolve 20 mg of naphthol AS-MX phosphate in 0.4 ml of *N*,*N*-dimethylformamide in a glass vessel (plastic vessels will dissolve). Add 19.6 ml of Tris/HCl quickly with stirring. This solution can be kept at 4 °C for several weeks or it can be stored frozen in 1 ml aliquots.

- Incubating solution
  - Substrate stock solution
  - Fast Red TR salt (Sigma F8764)
  - 0.1 M Tris/HCl buffer (pH 8.2)
  - Levamisole (Sigma L9756)
- Mayer's haemalum
  - 1 g of haematoxylin
  - 50 g of potassium or ammonium alum
  - 0.2 g of sodium iodate
  - 1 g of citric acid
  - 50 g of choral hydrate
  - 1 litre of distilled water

  Dissolve the haemotoxylin, potassium alum and sodium iodate in the distilled water overnight at room temperature. Add the citric acid and choral hydrate. Mix and boil for 5 min. Cool and filter

Procedure

1. Thaw a 1 ml aliquot of the substrate solution.
2. Weigh 5 mg of Fast Red TR salt and add it to 4 ml of 0.1 M Tris/HCl (pH 8.2) containing 0.1 mM levamisole (approximately 1.2 mg).
3. Just before use, add the 1 ml of substrate solution and shake well to dissolve as much as possible of the Fast Red TR salt.
4. Filter onto the preparations and incubate for 5–20 min either at room temperature or at 37 °C. Check the reaction microscopically until the red colour is satisfactory.
5. Wash the preparations well in running tap water.
6. Counterstain nuclei lightly in a progressive nuclear stain such as Mayer's haemalum (to avoid differentiation). Do not differentiate in acid–alcohol because the red end-product is alcohol-soluble. If you have to differentiate, use weak aqueous hydrochloric acid. Blue in tap water or an alkaline solution, rinse in water but do not dehydrate. Mount in an aqueous mountant such as Hydromount (National Diagnostics, HS-106)

or, for better microscopical resolution, Aquaperm (Shandon, 484975) or any similar mountant.

Method 9. Development of a purple/brown reaction product from alkaline phosphatase

This method uses a more expensive substrate, BCIP and a tetrazolium salt as chromogen. The reaction product is purple/brown and is less soluble in alcohol than the azo dyes of the reaction described in Method 8. It is said to be a more sensitive reaction, better able to reveal smaller quantities of alkaline phosphatase, and it is often used in conjunction with alkaline phosphatase-labelled probes for *in situ* hybridization of nucleic acids. Kits are available commercially.

Materials

- BCIP, disodium salt (Sigma B 6149)
- *N*,*N*-Dimethylformamide
- Tetranitro Blue Tetrazolium (TNBT) (Sigma T 4000)
- 0.2 M Tris–HCl (pH 9.5)
- Magnesium chloride ($MgCl_2 \cdot 6H_2O$)
- Levamisole
- Solution A
  - Dissolve 5 mg of BCIP in 0.1 ml of *N*,*N*-dimethylformamide.
- Solution B
  - Dissolve 10 mg of TNBT in 0.2 ml of *N*,*N*-dimethylformamide. Add 0.1 ml of 0.2 M Tris/HCl (pH 9.5).
- Solution C
  - This consists of 30 ml of 0.2 M Tris/HCl (pH 9.5) containing 10 mM magnesium chloride (61 mg of $MgCl_2 \cdot 6H_2O$) and 1 mM levamisole (7.2 mg).
- Mayer's carmalum
  - 2 g of carminic acid
  - 100 ml of 5% aqueous ammonium alum
  - 0.2 g of salycyclic acid or thymol
  - Distilled water

Add the carminic acid to 100 ml of ammonium alum solution. Dissolve by boiling for 1 h. After cooling, return to the original volume with distilled water. Add salycyclic acid or thymol and mix thorougly. Filter and store at 4 °C.

Procedure

1. Mix together solutions A (0.1 ml), B (0.3 ml) and C (30 ml).
2. Pass the mixture through a filter onto the preparations and incubate for 30 min or as long as is required at room temperature.
3. Rinse the preparations in water. A suitable pink nuclear counterstain is Mayer's carmalum or Nuclear Fast Red if the preparation is to be mounted in an aqueous mountant such as Hydromount or Aquaperm. If the preparation is dehydrated (this should be done rapidly because some reaction product will dissolve in alcohol), Methyl Green, Mayer's carmalum, Nuclear Fast Red or Neutral Red are suitable. Alternatively, the counterstaining step can be omitted.

5. Controls

Positive and negative controls are essential components of every run of every immunohistochemical reaction to ensure that the reaction is working properly and that the result is specific.

5.1. Positive controls

These are preparations known to contain the antigen being localized and prepared in the same way as the test tissue. If they do not show a positive reaction, something has gone wrong with the method.

5.2. Negative controls

These preparations are the same as the test tissue and are treated identically except for the omission of the primary antibody, which can be substituted with PBS (antibody diluent). An inappropriate antibody (of the same Ig subclass for a monoclonal antibody) is an alternative negative control. Normal serum from the same species as the primary antibody (at a standard dilution that does not to bind to control tissue) is a suitable negative control when using a polyclonal primary antibody. A reaction on the negative control preparation indicates non-specific binding of one or more reagents to the test tissue. Some possible causes are inadequate blocking of tissue binding sites, unblocked endogenous biotin, endogenous enzyme, or cross-reaction of the second antibody with Ig of the test tissue species (see Section 8).

If a new antibody of uncertain specificity is being investigated, pre-absorption of the diluted primary antibody with an excess of its purified antigen is an essential control to show that immunoreactivity is removed by the antigen.

6. Multiple immunostaining

It is sometimes necessary to immunostain more than one antigen in the same preparation; they are identified with different labels. Many of the methods for doing this are complicated, and beyond the scope of this chapter, particularly when the two or more primary antibodies are raised in the same species, because of the problems of cross-reaction between the various reagents. However, no special problems are associated with using two differently labelled primary antibodies together or sequentially in the direct method, regardless of the species in which they were raised. Fluorescent or enzyme labels can be used. Similarly, an indirect or three-layer method can be used without problem if the two primary antibodies are raised in different species and species-specific second antibodies are used [23]. PAP and APAAP can be combined or one of the second antibodies can be biotinylated and detected with labelled avidin. The enzyme activities are developed separately after all the immunoreagents have been applied. Full discussion of this area is provided by van der Loos [24].

7. Immunostaining combined with other methods

Any labelled reaction such as a histochemical enzyme reaction, a tissue stain or *in situ* hybridization for mRNA or DNA can be combined with an immunostain, provided that the reactions are compatible. Each situation will require individual attention, full understanding of the various stages and experiments to determine the best order of procedure. For example, the effect of the protease treatment required for *in situ* hybridization on the immunoreactivity of the antigen may have to be determined. If *in situ* hybridization follows an immunocytochemical method the investigator must ensure that the nucleic acid to be identified is not affected by, for example, heat-induced antigen retrieval. The end-products of combined reactions with enzyme labels must be easily distinguishable, e.g. the blue or purple/brown colour developed from alkaline phosphatase and the red colour developed from peroxidase with aminoethylcarbazole. It may be better to develop each enzyme separately after both sets of immunoreagents have been applied, in case the reaction product of the first reaction obscures the binding site of the second [24].

Sections of tissue fixed in methacarn [methanolic Carnoy's fixative (see Section 8.1.1)], even after paraffin embedding, can be used for the extraction of proteins for Western blotting. The adjacent sections can be immunostained for the same proteins, thus providing combined information on their structure and localization [25].

8. Troubleshooting

There are many reasons why an immunoreaction can go wrong. Just a few of the most common are listed here. Problems frequently result from human error; therefore check all reagents and the order in which they were applied, and ensure that the paraffin is completely removed from paraffin sections. Even a small error such as the wrong buffer pH can adversely affect an immunostain.

8.1. Too little staining

This may be due to a fault in the preparation method, staining method or reagents, or to lack of sensitivity of the method if the amount of antigen to be stained is very small.

8.1.1. Fixation

Fixation may have altered the immunoreactivity of the antigen, particularly if formaldehyde was used. Try washing the tissue thoroughly before processing and embedding to reverse some of the cross-linking effects of formalin. If this is not possible, use antigen retrieval methods (Methods 3–5) or try a precipitant fixative instead, e.g. methacarn [methanol/chloroform/glacial acetic acid (6:3:1 by volume)].

8.1.2. Permeabilization

This may be necessary to allow penetration of antibodies into whole cell preparations or thick, pre-fixed, frozen sections. If the preparations are formalin-fixed, further treatment with methanol may help. Provided that the antigenicity is not affected, thick sections or whole mounts should be dehydrated through graded alcohols to xylene and rehydrated [26]. For cell preparations, add a detergent such as saponin (0.5–1%) to the washing buffer.

8.1.3. Peroxidase blocking

This step can damage some antigens in frozen sections. Try immunostaining without the blocking step (compare with a negative control to see the extent of endogenous peroxidase), or block after incubating with the primary antibody but before the peroxidase-labelled step. If necessary, try a milder peroxidase blocking method [14].

8.1.4. Increasing sensitivity

(a) Build up layers of antibodies to get more label on the preparation. For example, after a rabbit PAP complex, apply a second layer of anti-rabbit Ig, and then another layer of rabbit PAP before developing with DAB/H_2O_2 (or even after the first development if the reaction is weak). Layers can also be repeated in the APAAP method.

> ***Note*** *Do not have azide in the diluent for the second layer of anti-rabbit Ig or you will block the peroxidase on the first layer of PAP.*

(b) Use an ABC method if you are not already doing so, or an immunogold reaction with silver enhancement, which is extremely sensitive and gives a black (metallic silver) end product [9,10].

(c) Combine ABC and PAP by using a high concentration of biotinylated second antibody so that one F_{ab} arm is free to bind to PAP in a third layer that is a mixture of PAP and ABC (which binds to the biotin), resulting in an increased quantity of label on the preparation.

(d) Increase the antigen retrieval time in enzyme or heated buffer.

Note *Too long an incubation in enzyme can damage the tissue structure; heated buffer is less damaging.*

(e) Use a catalysed signal amplification method. Peroxidase label (undeveloped) from the first reaction in the presence of H_2O_2 activates biotinylated tyramine. The tyramine binds to tissue proteins at the antigenic site. A further layer of labelled avidin or ABC (any label) becomes attached to the extra biotin, increasing the final amount of label on the preparation [27,28]. Kits are available commercially for this procedure or reagents can be made up in the laboratory. The dilution of the primary antibody can be increased with this method.

8.1.5. Enhancing the visibility of a weak peroxidase reaction

There are several ways to darken the product of the DAB/H_2O_2 reaction.

(a) After the reaction, immerse the preparation in 0.5% copper sulphate in 0.9% sodium chloride for approx. 2 min, then rinse and counterstain as usual. There are also commercially available DAB enhancing solutions which are usually silver or gold salts (e.g. Vector Laboratories H-2200).

(b) Another useful enhancing solution is hot methenamine-silver [29]. These reactions give a black product.

(c) The inclusion of 0.01 M imidazole (Sigma I 0125) in the developing solution darkens the brown colour of the reaction product [30].

(d) The inclusion of nickel or cobalt salts in the developing solution gives a dark blue/black product [31].

8.2. Too much staining

Structures other than the antigen may be stained so much that the specific immunostaining cannot be distinguished from the background. There are standard procedures to reduce background staining that results from non-specific binding of antibodies to tissue components.

(a) Dilute the antibodies further; this usually applies to the primary antibody, whether polyclonal or monoclonal, as the dilutions of the second and third reagents should have been optimized in previous reactions. Immunostain with a series of doubling dilutions of the primary antibody until the specific staining diminishes, then use the dilution one or two steps more concentrated to ensure saturation of the antigen.

(b) Reduce any non-specific binding of primary antibody: (a) by adding detergent to the washing buffer (e.g. 0.05% Tween 20, or 0.1% saponin for cell preparations to reduce leaching of membrane proteins); (b) by increasing the pH of the washing buffer and the antibody diluent to 9 or 10; (c) by increasing the sodium chloride content of the washing buffer from 0.9% to 2.5%.

(c) Increase the concentration of the blocking serum (even undiluted serum can be used) to saturate F_c receptors (fresh tissue) or hydrophobic or charged sites in the tissue substrate.

(d) Check that any unwanted staining is not due to endogenous peroxidase activity. If this appears in negative control preparations, blocking should be performed as described in Section 3.3.2. If staining is due to endogenous biotin, blocking should be carried out as in Method 1. Pigment in the tissue may masquerade as positive staining.

(e) If aldehydes remain in the tissue (particularly after fixation with glutaraldehyde, but also after long fixation in formalin), amino groups of antibodies can bind covalently to them. Aldehydes can be reduced to non-attracting alcohols by incubating in freshly prepared 0.02% sodium borohydride or potassium borohydride for 2 min [32] or by adding 50 mM ammonium chloride [33] to the blocking serum.

(f) Secondary antibodies (anti-Igs), can react with Igs of the host species as well as with those of the species supplying the primary antibody. Test for this by staining lymph nodes or spleen tissue of the host species with the secondary antibody. The test tissue must be prepared in the same way as the experimental tissue. If the Ig-containing plasma cells are stained, block the cross-reaction by adding 1–5% normal serum from the host species to the working dilution of the secondary antibody. The working concentration of the secondary antibody may have to be increased to compensate for the cross-reactive antibody population that has been removed by the added Ig. Species-specific (affinity-purified or pre-absorbed) secondary antibodies are available from most suppliers.

(g) Primary antibodies can cross-react with antigens that are similar molecules to the ones being investigated. If a polyclonal antibody is being used it may be possible to absorb out the cross-reacting population with the similar molecule, leaving a region-specific primary antibody. A monoclonal antibody cannot be corrected like this as it consists of only one type of antibody. In this case there is no cure except trying another antibody raised to a different, unshared epitope.

(h) Autofluorescence can be a problem in immunofluorescent preparations if it is of a similar colour to the fluorescent label. It can be disguised with a fluorescent counterstain such as Pontamine Sky Blue (0.05% in PBS/1% dimethyl sulphoxide) [34]. Apply the stain for 30 min before the application of the primary antibody. This stains fluoresces red with the filters used for fluorescein. Alternatively, use a label that fluoresces in a different colour.

References

1. Polak, J.M. and Van Noorden, S. (1997) Introduction to Immunocytochemistry, 2nd edn, Bios Scientific Publishers, Oxford

2. Beesley, J.E. (ed.) (1993) Immunocytochemistry: a Practical Approach, Oxford University Press, Oxford

3. Larsson, L.-I. (1988) Immunocytochemistry: Theory and Practice, CRC Press, Boca Raton, FL

4. Sternberger, L.A. (1986) Immunocytochemistry, 3rd edn, John Wiley & Sons, New York

5. Polak, J.M. and Priestley, J.V. (1992) Electron Microscopic Immunocytochemistry: Principles and Practice, Oxford University Press, Oxford

6. Coons, A.H. and Kaplan, M.H. (1950) J. Exp. Med. **91**, 1–13

7. Ritter, M.A. and Ladyman, H. (eds) (1995) Monoclonal Antibodies, Production, Engineering and Clinical Application, Cambridge University Press, Cambridge

8. Johnstone, A. and Thorpe, R. (1996) Immunochemistry in Practice, 3rd edn, Blackwell Scientific Publications, Oxford

9. Holgate, C.S., Jackson, P., Cowen, P.N. and Bird, C.C. (1983) J. Histochem. Cytochem. **31**, 938–944

10. Hacker, G.W., Grimelius, L., Danscher, G., Bernatzky, G., Muss, W., Adam, H. and Thurner, J. (1988) J. Histotechnol. **11**, 213–221

11. Wood, G.S. and Warnke, R. (1981) J. Histochem. Cytochem. **29**, 1196–1204

12. Sternberger, L.A., Hardy, Jr, P.H., Cuculis, I.J. and Mayer, H.G. (1970) J. Histochem. Cytochem. **18**, 315–333

13. Hsu, S.M., Raine, L. and Fanger, H. (1981) J. Histochem. Cytochem. **29**, 577–580

14. Andrew, S. and Jasani, B. (1987) Histochem. J. **19**, 426–430

15. Shi, S.-R., Key, M.E. and Kalsa, K.L. (1991) J. Histochem. Cytochem. **39**, 741–748

16. Cattoretti, G., Pileri, S., Parravicini, C., Becker, M.H.G., Poggi, S., Bifulco, C., Key, G., D'Amato, L., Sabattini, E., Feudale, E. et al. (1993) J. Pathol. **171**, 83–98

17. Norton, A.J., Jordan, S. and Yeomans, P. (1994) J. Pathol. **173**, 371–379

18. Morgan, J.M., Navabi, H., Schmid, K.W. and Jasani, B. (1994) J. Pathol. **174**, 301–307

19. Shi, S-R., Imam, A., Young, L., Cote, R.J. and Taylor, C.R. (1995) J. Histochem. Cytochem. **43**, 193–201

20. Morgan, J.M. Navabi, H. and Jasani, B. (1997) J. Pathol. **182**, 233–237

21. Shi, S-R., Cote, R.J. and Taylor, C.R. (1997) J. Histochem. Cytochem. **45**, 327–343

22. Pelliniemi, L.J., Dym, M. and Karnovsky, M.J. (1980) J. Histochem. Cytochem. **28**, 191–192

23. Mason, D.Y. and Sammons, R.E. (1978) J. Clin. Pathol. **31**, 454–462

24. van der Loos, C.M. (1999) Immunoenzyme Multiple Staining Methods, Bios Scientific Publishers, Oxford

25. Lalani, E.-N., Golding, M., Hudson, M., Chieffi, G., Stamp, G., Anilkumar, T.V., Sarraf, C. and Alison, M.R. (1995) J. Pathol. **177**, 323–328

26. Costa, M., Buffa, R., Furness, J.B. and Solcia, E. (1980) Histochemistry **3**, 157–165

27. Adams, J.C. (1992) J. Histochem. Cytochem. **40**, 1457–1463

28. Merz, H., Malisius, R., Mannweiler, S., Zhou, R., Hartmann, W., Orscheschek. K., Moubayed, P. and Feller, A.C. (1995) Lab. Invest. **73**, 149–156

29. Peacock, C.S., Thompson, I.W. and Van Noorden, S. (1991) J. Clin. Pathol. **44**, 756–758

30. Straus, W. (1982) J. Histochem. Cytochem. **30**, 491–493

31. Shu, S., Ju, G. and Fan, L. (1988) J. Histochem. Cytochem. **85**, 169–171

32. Craig, A.S. (1974) Histochemistry **42**, 141–144

33. Tougard, C. and Picart, R. (1986) Am. J. Anat. **175**, 161–167

34. Cowen, T., Hven, A.J. and Burnstock, G. (1985) Histochemistry **82**, 205–208

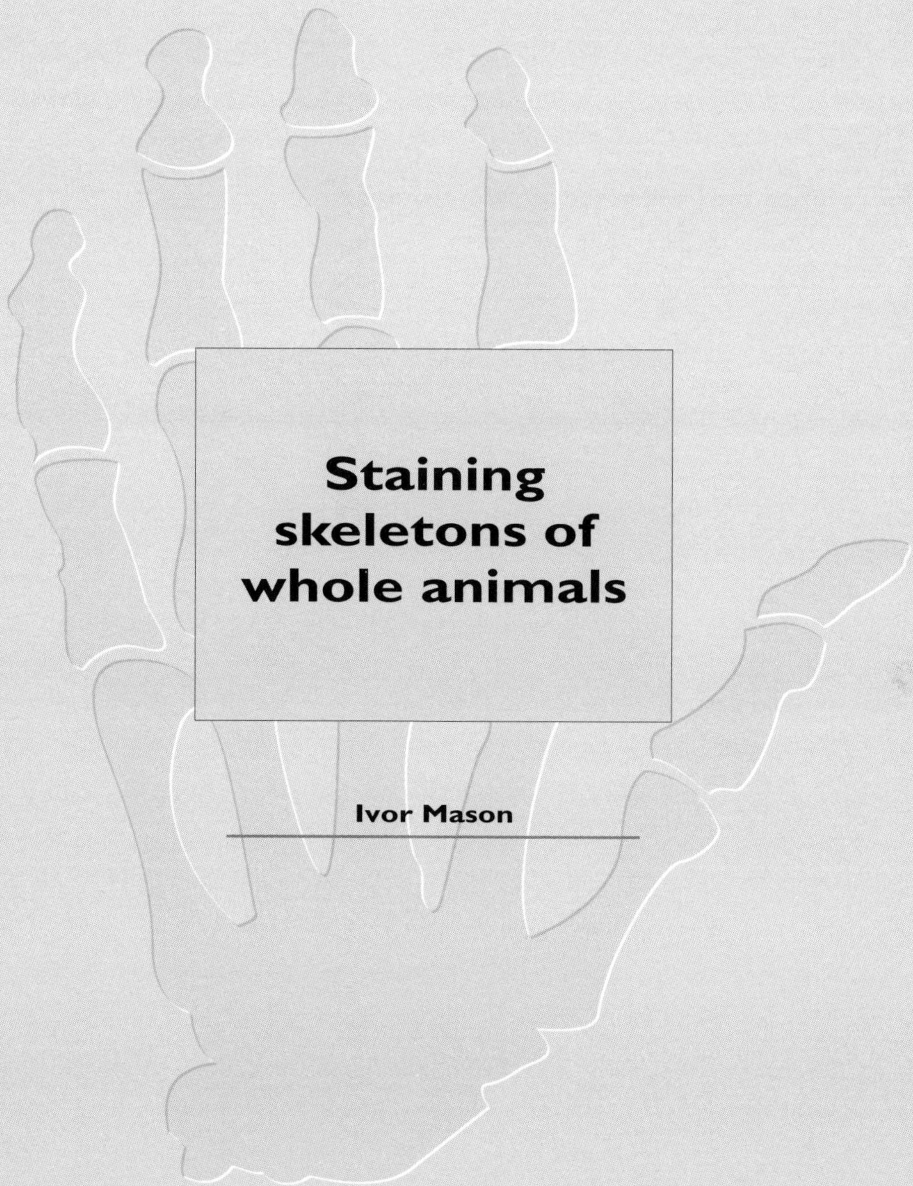

Staining skeletons of whole animals

Ivor Mason

Chapter 10

1. Introduction

The staining of vertebrate skeletal material has proven a very informative technique in the analysis of mutant specimens or those undergoing pathological processes, e.g. in response to teratogenic processes. Indeed, many transgenic 'knockout' mouse strains show skeletal defects that are readily identified by the following simple staining protocols. The first procedure details a simple stain for cartilage. The two subsequent procedures describe methods of staining and distinguishing both bone and cartilage.

Method 1. Staining for cartilage in embryos

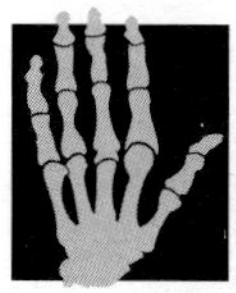

The following procedure stains cartilage blue–green while leaving other tissues optically transparent (see Fig. 1A).

Materials

- Bouin's fixative
  - See Chapter 6, section 6.2.3, for composition.
- 1% (v/v) Ammonia (Merck BDH, Poole, U.K.) in 70% (v/v) ethanol (Merck BDH)
- 5% (v/v) Acetic acid (Merck BDH)
- 1% (w/v) Alcian Blue 8GX (Sigma-Aldrich, Poole, U.K.) in 5% (v/v) acetic acid
- Graded ethanols (70%, 95% and 100%)
- Benzyl benzoate/methyl salicylate (1:3) (both Merck BDH)

Procedure

1. Fix the specimen in Bouin's fixative for 4 h to overnight, depending on its size.

2. Wash in several changes of distilled water.

3. Wash four times in 1% (v/v) ammonia in 70% (v/v) ethanol (1 h for each wash).
4. Mix 95 ml of 5% (v/v) acetic acid with 5 ml of 1% (w/v) Alcian Blue 8GX in 5% (v/v) acetic acid. Incubate each embryo in 25–50 ml of this solution for 4 h to overnight, depending on size.
5. Wash three times [twice for 1 h and once overnight (minimum 16 h)] in 5% acetic acid (50 ml per embryo).
6. Dehydrate in graded ethanols — 1 h in 70% ethanol, 1 h in 95% ethanol and three times in 100% ethanol for 1 h each (50 ml per embryo).
7. Clear and store the material in benzyl benzoate/methyl salicylate (10 ml per embryo; embryos can eventually be stored in a smaller volume).

2. Staining for cartilage and bone in the same preparation

In the following procedures cartilage stains blue–green and mineralized bone stains red (see Figs. 1B–1E).

Method 2. Staining cartilage and bone embryonic material

Materials

- 95% (v/v) Ethanol
- Staining solution
    - Alcian Blue 8GX
    - Glacial acetic acid/95% (v/v) ethanol (1:4, (v/v))

  Dissolve 15 mg of Alcian Blue 8GX in 100 ml of glacial acetic acid/95% (v/v) ethanol (1:4, v/v).
- 1% (w/v) Potassium hydroxide (KOH)
- 0.1% (w/v) Alizarin Red S (freshly prepared and containing 500 μl of 1% (w/v) KOH per 100 ml
- 20% (v/v) Glycerine (Sigma-Aldrich) in 1% KOH
- 70% Ethanol/glycerine/water

These should be in the following proportions: 1:2:7, 2:2:6, 3:3:4, 4:4:2 and 5:5:0 (by vol.).

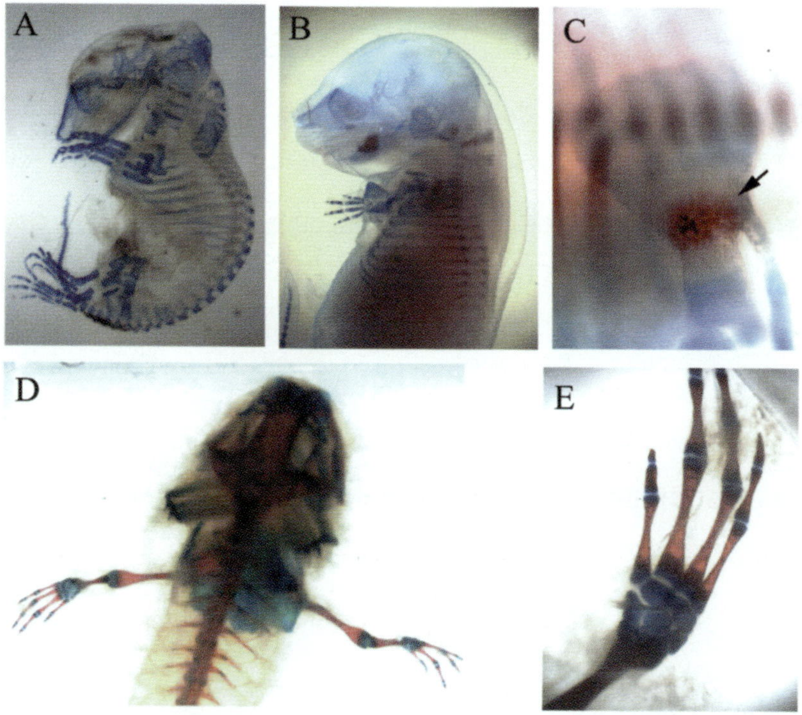

Fig. 1. Staining embryos and adult skeletons for bone and cartilage

(A) Cartilage formation in a 12.5-day mouse embryo. (B) A 17-day mouse embryo, stained for cartilage (blue) and bone (red) formation, showing the onset of ossification in the ribs, lower jaw and scapula. (C) Higher magnification of the scapula of the embryo in (B) to show the onset of ossification (arrow). (D) Staining of bone and cartilage in an adult axolotl (Ambystoma). (E) Higher magnification of the footplate of the forelimb of the specimen in (D). The axolotl skeletal preparation was produced by my colleague Malcolm Maden.

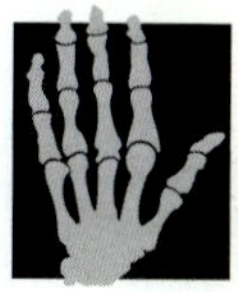

Procedure

1. For best results, remove skin.
2. Fix in 95% ethanol.
3. Stain in a large volume of the staining solution.
4. Differentiate the colour in three changes of 95% ethanol over a period of 1 week.
5. Place in 1% KOH until the bones become visible.
6. Stain in Alizarin Red S for 3–12 h.

Check for red colour; it may be necessary to add a little acid to obtain this.

7. Wash in running tap water for 30 min.
8. Decolourize for 1–2 weeks in 20% glycerine (in KOH).
9. Dehydrate through the graded ethanol/glycerine/water from the 1:2:7 ratio through to the 5:5:0 ratio. This will clear the specimen and they can be stored in the final solution.

Method 3. Staining cartilage and bone postnatal skeletons

Materials

- Ethanol
- Acetone
- Staining solution
  - 0.3% (w/v) Alcian Blue 8GX in 70% ethanol
  - 0.1% (w/v) Alizarin Red S in 95% ethanol
  - 100% acetic acid
  - Autoclaved distilled water

  These should be mixed in the proportions of 1:1:1:17 (by vol.) and the mixture should be prepared on day of use.
- 20% (v/v) Glycerol/2% (w/v) KOH solution
- 50%, 80% and 100% (v/v) Glycerol
- Phosphate-buffered saline

Procedure

1. Fix the specimen in ethanol for 4 days.

   Specimens that have already been fixed in formaldehyde for days or months can also be used. They should be washed during the day and then overnight in several changes of 1× phosphate-buffered saline and then ethanol-fixed as above.

2. Wash in several changes of acetone, each for 1 h, and then leave for 3 days in acetone.
3. Rinse in distilled water.
4. Place in a large volume of staining solution for at least 10 days.
5. Rinse in distilled water.
6. Incubate in 20% (v/v) glycerol/2% (w/v) KOH solution overnight at 37 °C.
7. Keep specimens incubated in this solution at 37 °C for 2–3 weeks, changing the solution every 2–3 days until the specimens have completely cleared.
8. Incubate in 50%, 80% and 100% glycerol (for 2 days each) and store in 100% glycerol.

I am grateful to Gillian Morriss-Kay, who introduced these procedures to my group. Work in the author's laboratory is supported by programme grants from the Medical Research Council and the Human Frontier Science Program, and project grants from the Biotechnology and Biological Sciences Research Council and the Wellcome Trust.

Metabolic mapping by enzyme histochemistry

Cornelis J.F. Van Noorden*
and Wilma M. Frederiks

*To whom correspondence should be addressed.

Chapter 11

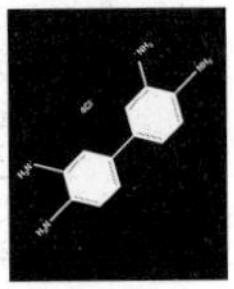

1. Introduction

Enzyme histochemistry is the discipline that is applied to detect and localize activities of enzymes in cells and tissues. It is used for the determination of local metabolism in individual cells or tissue compartments, or, in other words, for metabolic mapping. Enzyme activities in intact cells or tissues often differ greatly from the activities exhibited by enzymes in diluted cell-free systems, owing to complex interactions between the enzymes and other macromolecules in their environment *in vivo*. These interactions are lost in cell-free systems. Therefore, in order to understand the role of an enzyme in a living organism, it is essential that the enzyme be studied in intact cells and tissues. A recent closer approximation to *in vivo* reality is the use of enzyme histochemical methods to map metabolic activity directly in living cells [1,2]. In recent years, important new methodologies to detect the activity of hydrolases and, in particular, of proteases have been developed. These are based on the hydrolysis of specific fluorogenic substrates, which generates fluorescent products. These methodologies have been reviewed in [2].

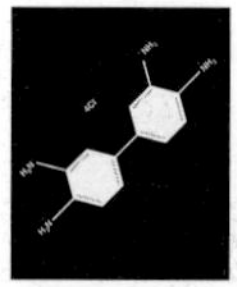

An example of the complexity of regulation of enzyme activity in intact tissues is shown in Fig. 1. The localization pattern of activity of the proteolytic enzyme cathepsin B is directly compared with the immunolocalization of the enzyme in adjacent sections of human colon. Although cathepsin B protein is localized ubiquitously in colon mucosa, the enzyme is active only in certain regions. This example shows how important it is to keep cells and tissues intact for understanding how enzymes function *in vivo*, and also that activity is a more accurate measure of the physiological function than the presence of an enzyme protein. Enzyme histochemistry will become an even more valuable tool in the 21st Century than it has been so far. As the complete genomes of several organisms are mapped, the critical next step will be to clarify the specific functions of the proteins encoded by the genes. Thus, after the era of genome mapping, the focus of research in the life sciences will shift to metabolic mapping.

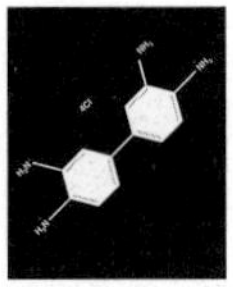

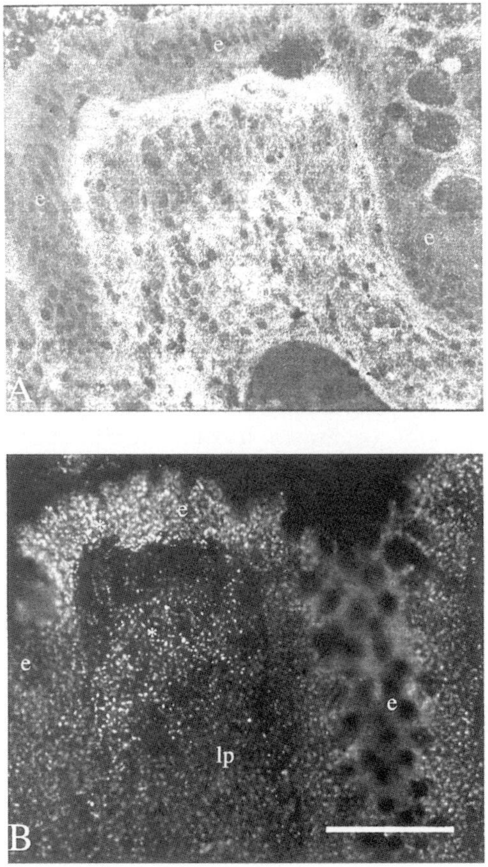

Fig. 1. Localization of the proteinase cathepsin B in adjacent sections of human colon

(A) Immunohistochemical localization of cathepsin B protein using an immunogold method and silver enhancement. Cathepsin B protein is represented by white dots as demonstrated with reflection contrast microscopy. (Photomicrograph by Dr A. Köhler, Alfred Wegener Institute for Marine and Polar Research, Bremerhaven, Germany.) (B) Localization of cathepsin B activity using the fluorescence method described in Method 15. The white dots represent activity. Bar = 50 μm. The epithelium (e) contains relatively low levels of cathepsin B molecules, which are active only at the top of the crypts whereas the lamina propria (lp) contains high levels of cathepsin B that is active in a distinct area only.

2. Principle of enzyme histochemical methods

Enzyme histochemical methods are based on the principle that the enzyme of interest in a cell preparation or tissue section converts its substrate to a coloured or fluorescent product that precipitates at the site of the activity. Coloured or fluorescent product formation can be observed with a microscope during incubation (Fig. 2) or after the reaction has been stopped by rinsing off the incubation medium and mounting the cells or sections (Fig. 3). In this way, metabolic activity of individual cells in a population or in a tissue can be determined. In general, the incubation medium is aqueous and the substrate of the enzyme is water-soluble, whereas the coloured or fluorescent product is water-insoluble and, therefore, it precipitates at the site where it is formed.

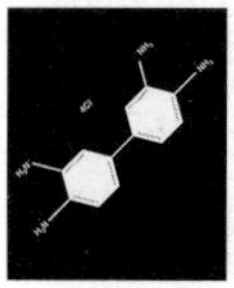

The extent of product formation can be quantified by image analysis. In this case, the coloured or fluorescent image is observed in the microscope and is captured by an attached video or charge-coupled device camera, which converts the microscopic image into a digital image (Fig. 2). Digital images can then be analysed by computer to obtain quantitative data [3]. In this way, quantitative information about metabolic activity in individual cells can be combined with morphological information (Fig. 4). Furthermore, by varying the substrate concentration in the incubation medium for a series of cell preparations or serial sections, kinetic parameters can be determined for the enzyme in its physiological environment (Fig. 4).

At present, the activities of only a limited number of enzymes can be localized, because not enough valid histochemical methods are available. Technological investments are therefore required in the near future to expand the enzyme histochemist's arsenal. Most of the valid methods that are available are summarized in this chapter. More thorough handbooks that are available [4–11] provide complete overviews of all methods that have been developed so far.

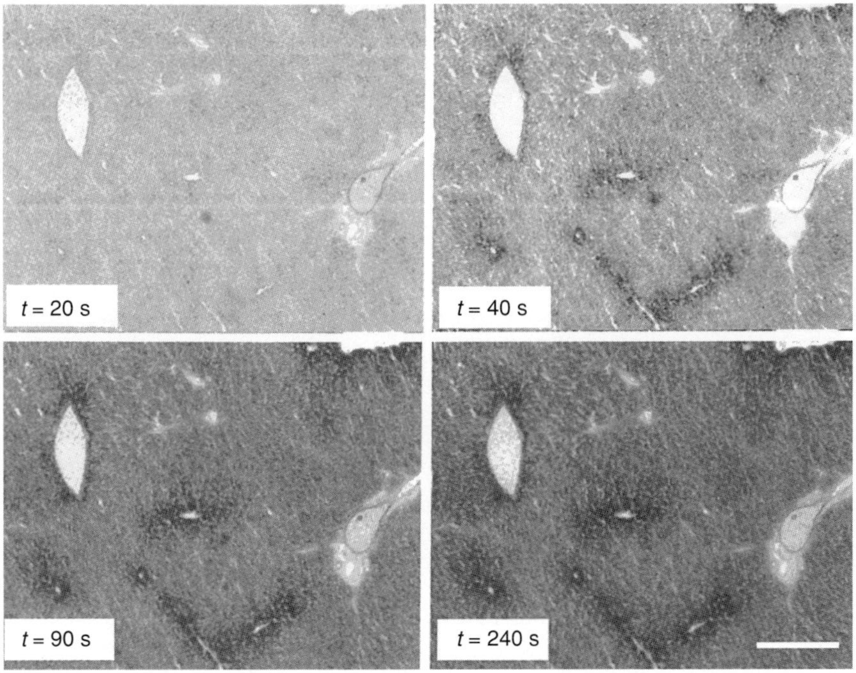

Fig. 2. Digital images of a rat liver section that has been incubated to localize mitochondrial glutamate dehydrogenase activity using Method 5

Images were captured at 20, 40, 90 and 240 s after the incubation was started by pouring the medium onto the section. Bar = 50 µm. In this way, the enzyme reaction can be followed quantitatively in time ([3]; see also Fig. 4).

3. Cell and tissue preparation for enzyme histochemistry

Most enzymes are rapidly inactivated by chemical fixation or embedding. Therefore, unfixed cells and tissues must usually be used for enzyme histochemical analysis. Freezing of unfixed cells and tissues in liquid nitrogen, storage at −80 °C and cryostat sectioning are the preparative methods of choice for enzyme histochemistry.

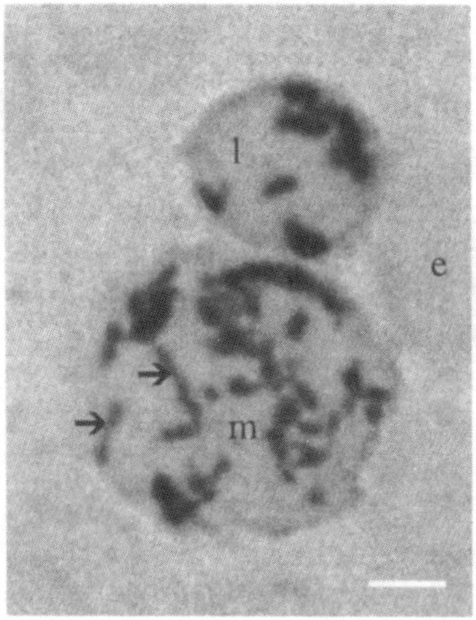

Fig. 3. Photomicrograph of unfixed human blood cells after incubation for a fixed time (5 min) to localize mitochondrial glutamate dehydrogenase activity using Method 5

After incubation, the medium was rinsed off and the cells were mounted in glycerol/gelatin. Activity was present only in the mitochondria (arrows) of white blood cells [a monocyte (m) and a lymphocyte (l)], but not in erythrocytes (e), which lack mitochondria. Bar = 5 μm.

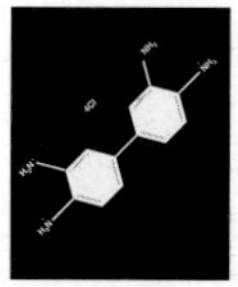

Enzyme histochemistry on individual cells is performed on cell smears or cytospin preparations after air drying. Method 1 describes the preparation method for unfixed cells. When preparations are handled according to this method, and the enzyme incubations are performed carefully (see below), excellent morphology can be retained without loss of enzymic activity. Localization of the product can be so precise that individual organelles can be distinguished (Fig. 3). Sufficient permeability of cell membranes is achieved by drying cell preparations, but an extra freeze–thawing step enhances permeability.

Preparation of tissues for enzyme histochemistry is performed as described in Method 2.

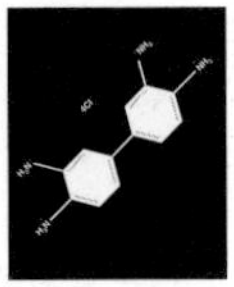

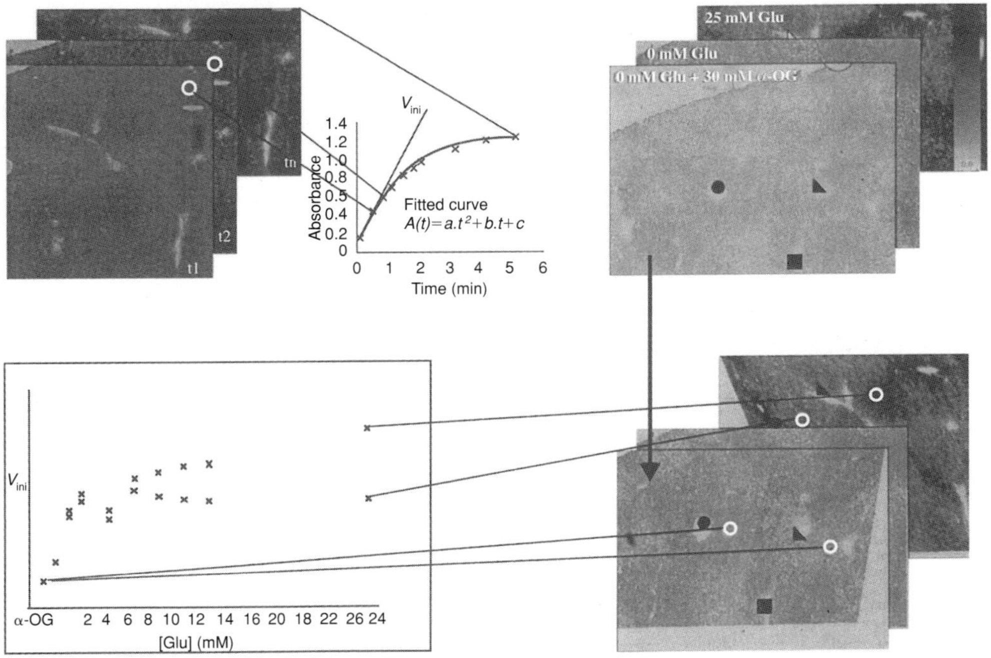

Fig. 4. Schematic representation of quantitative enzyme histochemical methodology

Activity of hepatic mitochondrial glutamate dehydrogenase is localized in sections as shown using Method 5. The upper left panel shows images that were captured at different times during incubation (as in Fig. 2). The upper middle plot shows the non-linear reaction in time in one area of the section that is described by the best fitting curve $A(t) = at^2 + bt + c$, where A is the absorbance, t is time, and a, b and c are constants. The initial velocity of the enzyme-catalysed reaction (V_{ini}) is represented by the constant b. The right-hand panels show images of serial sections that were incubated in various concentrations of glutamate (Glu) and a specific inhibitor of glutamate dehydrogenase, α-oxoglutarate (α-OG). The upper right panel shows images that were captured at the beginning of the reaction, while the lower right panel shows images at the end of the reaction. From these images, V_{ini} was calculated for each concentration of glutamate. The lower left plot shows the initial velocity (V_{ini}) versus glutamate concentration. This graph can be used to calculate the V_{max} and K_m of the enzyme [3]. V_{max} is the maximum velocity of the enzyme at optimum substrate concentration. K_m is the substrate concentration that gives half of the enzyme's maximum velocity. V_{max} is a parameter for the number of active enzyme molecules, whereas K_m is a parameter for the affinity of an enzyme for its substrate. A high K_m means low affinity and vice versa.

Sectioning is another critical step, especially when unfixed frozen material is used. Sections are cut with a cryostat as follows. Frozen tissue blocks that are kept at −80 °C are equilibrated with the temperature of the cryostat cabinet. Cut sections and pick them up on glass slides according to Method 3. Thawing and freezing are detrimental to morphology because of formation of ice crystals. Care must therefore be taken to keep the sections frozen until they are used. For extensive troubleshooting in relation to tissue sectioning, see Van Noorden and Frederiks [4].

Method 1. Unfixed cell preparations

Materials

- Earle's balanced salt solution
- Serum albumin
- Glass slides
- Centrifuge (cytospin or standard with special buckets)
- Desiccator

Procedure

1. Centrifuge cell suspension for 10 min at 4 °C (50–2000 *g*, depending on the cell type).
2. Resuspend cells in an Earle's balanced salt solution containing 5% (w/v) serum albumin.
3. Use 50 μl of cell suspension with approx. 10^6 cells/ml for cytospin preparations. Deposit cells on glass slides using a cytospin centrifuge or special buckets in a standard centrifuge. Centrifuge for 5 min at 400 × *g* at room temperature.
4. Dry the preparations in a desiccator, and store at −80 °C.

Method 2. Snap freezing of tissues

Materials

- Small vented plastic vials
- Liquid nitrogen
- Razor blade

Procedure

1. As soon as the tissue is obtained, cut it with the razor blade into 5 mm thick blocks.
2. Put the pieces of tissue into small vented plastic vials and close the vials.
3. Slowly immerse the vials containing the piece of tissue in liquid nitrogen and keep them there for at least 5 min.
4. Store the vials with tissue in a biological freezer at −80 °C.

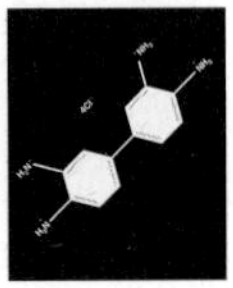

Method 3. Cutting unfixed cryostat sections

Materials

- Cryostat
- Glass slide
- TissueTek (OCT Compound; Sakura, Tokyo, Japan)

Procedure

1. Attach the tissue block to a metal chuck in the cryostat cabinet using TissueTek at an ambient temperature between −30 and −20 °C. Prevent warming of the tissue block by the liquid TissueTek by using a limited amount and letting it cool for a while before sticking the tissue to it.
2. Start cutting sections when the block has been trimmed to the desired level. The optimal thickness for cryostat sections is 8–10 μm but sections can be cut at any thickness between 2 and 30 μm.

3. Cut sections with the desired thickness at a slow but constant speed. When a section is correctly cut it remains flat on the microtome knife, under the anti-roll plate.
4. Pick up the section by holding a glass slide close to the section, which will then neatly jump onto the slide due to static electrical forces [6].
5. Store the sections at between −20 and −30 °C until they are to be used.

4. Enzyme histochemical methods

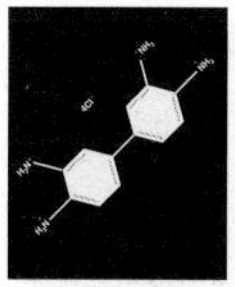

Incubation procedures for the demonstration of enzyme activities can be performed in various ways and the choice of procedure depends on the properties of the enzyme of interest and/or the principle of the histochemical method. All these incubation procedures are discussed at length in handbooks [4–11], for both light and electron microscopy. We limit ourselves here to a method that can be used for most enzymes and subsequent light microscopy. This procedure is based on the addition of a tissue protectant, polyvinyl alcohol, to the incubation medium. Addition of polyvinyl alcohol enables the localization of activity in unfixed cell preparations or cryostat sections without loss of enzymes and other large molecules due to diffusion. When polyvinyl alcohol is dissolved in water or buffer, the solution becomes viscous, but it is not the viscosity that gives the protective effect. Rather, protection is achieved by the principle of the excluded volume: water molecules are trapped by the polymer molecules so that less water is available for solvation and diffusion of larger molecules from cells or sections. In this way, all enzymes are kept in place and excellent morphology is retained (Figs. 1B and 3).

Not all grades of polyvinyl alcohol yield reliable results. Only polyvinyl alcohol with a number average molecular weight of 20 000–40 000 is suitable for use as a tissue protectant. Polyvinyl alcohol from Sigma (St Louis, MO, U.S.A.) with a number average molecular weight of 40 000 is the tissue protectant of choice for most enzyme histochemical reactions. The preparation of media containing polyvinyl alcohol is described in Method 4.

Method 4. Preparation of polyvinyl alcohol-containing incubation media

Materials

- Polyvinyl alcohol (Sigma; number average molecular weight, 40 000)
- Buffer
  - □ The type of buffer and the optimal concentration of polyvinyl alcohol vary with techniques for different enzymes, and are stated for each of the subsequent methods.
- Air-tight vials
- Spatula

Procedure

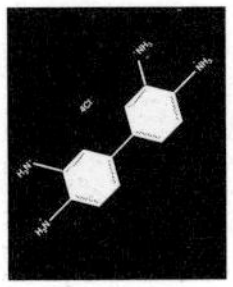

1. Dissolve 2–18 g of polyvinyl alcohol in 100 ml of buffer, with constant stirring and heating in a water bath until a clear solution is obtained.
2. Store the clear solution at 60 °C in air-tight vials.
3. Cool the desired volume of the solution to 37 °C before incubation.
4. Prepare the incubation medium by adding to the polyvinyl alcohol-containing buffer solution at 37°C the appropriate substrates and other compounds in dissolved form as concentrated stock solutions. Add 10 µl of a concentrated stock solution containing 100× the final concentration to 1 ml of buffer.
5. Mix the compounds thoroughly with the buffered polyvinyl alcohol solution using a spatula after each addition, because of the viscosity of the medium.
6. Incubate tissue sections or cell preparations in a drop (100–200 µl) of medium.

   When the medium contains a high concentration of polyvinyl alcohol (such as 18%) the drop remains by itself on the section or cell preparation, due to its viscosity. When the medium contains lower concentrations or no polyvinyl alcohol at all, use a PAP-pen (Dako, Glostrup, Denmark) to encircle the sections or cells with a water-repellant ring, as when performing immunohistochemical procedures.

Polyvinyl alcohol cannot be used in all conditions. For example, it can quench fluorescence and it can hamper penetration of compounds from the incubation medium into cells or tissues. In those cases, media have to be used with reduced concentrations of polyvinyl alcohol or even without this ingredient. The activity of enzymes that are tightly bound to structural elements of cells can usually be localized properly under these conditions.

Control incubations in the absence of substrate and in the presence of an inhibitor of the enzyme being studied are essential to establish specificity of the histochemical reaction. In general, inhibitors are used at the same molar concentrations as the substrate. Zollner [12] provides a useful guide for selecting specific inhibitors of enzymes.

Recommended concentrations of substrates and other components of the incubation medium are given for all methods. These concentrations are not necessarily optimal for all tissues and under all circumstances. Therefore, we recommend checking this by varying the concentrations for each tissue.

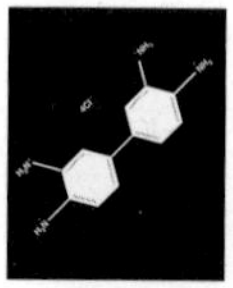

4.1. Oxidoreductases

Oxidoreductases are enzymes that oxidize and reduce substrates by transferring electrons from one substrate to another. The localization of dehydrogenase activity with the use of a tetrazolium salt is described in Method 5. A dehydrogenase oxidizes its substrate and transfers the removed electrons to its second substrate, which is either a coenzyme such as NAD^+ or $NADP^+$, or another electron acceptor. The activities of all dehydrogenases can be demonstrated with the tetrazolium salt method by adding the specific substrates and/or coenzymes to the incubation medium (see Table 1).

Localization of reductase activity is described in Method 6. This technique is widely used because of the association of NADPH–diaphorase activity with nitric oxide synthase. The specificity and validity of this histochemical method have recently been reviewed critically [13].

Two methods are described for the demonstration of oxidase activities. Oxidases are oxidoreductases that transfer electrons from their substrates directly to oxygen. A tetrazolium salt can replace oxygen, so oxidase activity can be demonstrated with the tetrazolium salt method as described in Method 7. Oxidase activity can also be demonstrated with the use of cerium, as described in Method 8.

Finally, methods are described to localize activities of catalase (Method 9) and peroxidase (Method 10). These enzymes are both important in cells for scavenging the reactive oxygen species hydrogen peroxide.

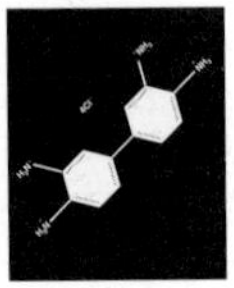

Method 5. Tetrazolium salt method for the demonstration of dehydrogenase activity

Materials

- Polyvinyl alcohol
- 0.1 M Phosphate buffer (pH 5.3)
- 0.1 M Phosphate buffer (pH 7.5)
- Substrate solution (see Table 1 for specific substrates and concentrations)
- Coenzyme solution (see Table 1 for specific coenzymes and concentrations)
- Sodium azide solution (33 mg/ml in distilled water)
- 1-Methoxyphenazine methosulphate solution (15 mg/ml in distilled water)
- Phenazine methosulphate solution (8 mg/ml in distilled water)
- Tetranitro Blue Tetrazolium (TNBT) or Nitro-Blue Tetrazolium (NBT) solution
  - Dissolve 5 mg in 20 μl of ethanol and 20 μl of dimethylformamide by heating gently.
- Glycerol/gelatin solution [50% (v/v) glycerol/7% (w/v) gelatin]

Table 1. Recommended concentrations of substrates and coenzymes for histochemical incubation media to demonstrate the activity of dehydrogenases with the tetrazolium salt method

Enzyme	Substrate(s)	Coenzyme
Glycerol-3-phosphate dehydrogenase (NAD^+-dependent) (EC 1.1.1.8)	50–100 mM Glycerol 3-phosphate	1.3 mM NAD^+
UDP–glucose dehydrogenase (EC 1.1.1.22)	3 mM UDP-glucose	3 mM NAD^+
Lactate dehydrogenase (EC 1.1.1.27)	120–150 mM Sodium lactate	3 mM NAD^+
3-Hydroxybutyrate dehydrogenase (EC 1.1.1.30)	100 mM D-3-Hydroxybutyrate	3 mM NAD^+
3-Hydroxyacyl-CoA dehydrogenase (EC 1.1.1.35)	100 mM β-Hydroxybutyryl-*N*-acetylcystamine or 20 mM β-hydroxy butyryl-coA	1 mM NAD^+
Malate dehydrogenase (NAD^+-dependent) (EC 1.1.1.37)	300 mM L-Malate	3 mM NAD^+
Malate dehydrogenase or malic enzyme ($NADP^+$-dependent) (EC 1.1.1.40)	100 mM L-Malate	0.8 mM $NADP^+$
Isocitrate dehydrogenase (NAD^+-dependent) (EC 1.1.1.41)	100 mM D,L-Isocitrate	7 mM NAD^+
Isocitrate dehydrogenase ($NADP^+$-dependent) (EC 1.1.1.42)	100 mM D,L-Isocitrate	4 mM $NADP^+$
Phosphogluconate dehydrogenase (EC 1.1.1.44)	8 mM 6-Phosphogluconic acid	0.8 mM $NADP^+$
Glucose-6-phosphate dehydrogenase (EC 1.1.1.49)	10 mM Glucose 6-phosphate	0.8 mM $NADP^+$
20α-Hydroxysteroid dehydrogenase (EC 1.1.1.62)	1.5 mM 20α-Hydroxy-4-pregnen-3-one	0.5 mM $NADP^+$
3β-Hydroxy-Δ^5-steroid dehydrogenase (EC 1.1.1.145)	2 mM Dehydroxyepiandrosterone (dissolved in dimethylformamide)	2 mM NAD^+ or 2 mM $NADP^+$
Glycerol-3-phosphate dehydrogenase (coenzyme-independent) (EC 1.1.99.5)	9.3 mM Glycerol 3-phosphate	–
Aldehyde dehydrogenase (EC 1.2.1.3)	25 mM Benzaldehyde, 6 mM acetaldehyde, 1.5 mM glutaraldehyde	1.5 mM NAD^+
Benzaldehyde dehydrogenase (EC 1.2.1.7)	25 mM Benzaldehyde, 6 mM acetaldehyde, 1.5 mM glutaraldehyde	1.5 mM $NADP^+$
Glyceraldehyde-3-phosphate dehydrogenase (EC 1.2.1.12)	2.3 mM Diethyl acetal salt of glyceraldehyde-3-phosphate	3 mM NAD^+
Succinate–semialdehyde dehydrogenase (EC 1.2.1.24)	50 mM 4-Aminobutyric acid, 30 mM α-oxoglutaric acid	1 mM NAD^+
Succinate dehydrogenase (EC 1.3.99.1)	50 mM Succinate	–
Glutamate dehydrogenase (NAD^+-dependent) (EC 1.4.1.2)	75 mM L-Glutamate	3 mM NAD^+

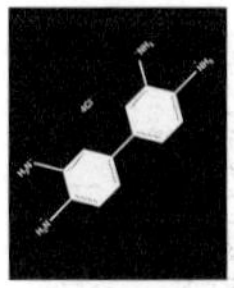

Procedure

1. Use unfixed cell preparations or cryostat sections. Let them equilibrate and dry for 5 min at 37 °C.
2. Dissolve 18 g of polyvinyl alcohol in 100 ml of 100 mM phosphate buffer, as in Method 4.
3. Add to 1 ml of polyvinyl alcohol-containing buffer [10 μl of 100× concentrated stock solutions of substrate and coenzyme (from Table 1)].
4. Add 10 μl of sodium azide solution (final concentration 5 mM).
5. Add either 10 μl of 1-methoxyphenazine methosulphate solution (final concentration 0.45 mM) or 10 μl of phenazine methosulphate solution (final concentration 0.2 mM).
6. Add 40 μl of TNBT or NBT solution (final concentration 5 mM).
7. Incubate for 5–30 min at 37 °C.
8. Rinse thoroughly with 0.1 M phosphate buffer (pH 5.3) at 60°C.
9. Mount in glycerol/gelatin.

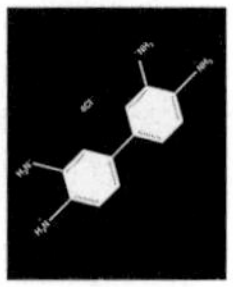

Control incubations should be carried out (i) in the absence of substrate or in the absence of substrate and coenzyme; (ii) in the presence of an inhibitor.

Method 6. Tetrazolium salt method for the demonstration of activity of NADPH–diaphorase (EC 1.6.99.–)

Materials

- Freshly depolymerized paraformaldehyde [4% in 0.1 M phosphate buffer (pH 7.4)]
- 30% Sucrose in 0.1 M phosphate buffer (pH 7.4)
- 0.3% Triton X-100 in 0.1 M phosphate buffer (pH 8.5)
- NADPH solution (84 mg/ml in distilled water)
- NBT solution
  - 1 mg dissolved in 20 μl of ethanol and 20 μl of dimethylformamide by heating gently.
- 0.1 M Phosphate buffer (pH 5.3)

- Glycerol/gelatin solution [50% (v/v) glycerol/7% (w/v) gelatin]
- 5 mM Dicumarol solution (160 mg/ml in distilled water)
- $NADP^+$ solution (400 mg/ml in distilled water)

Procedure

1. Fix cells or tissues in freshly depolymerized paraformaldehyde for 2 h at 4 °C.
2. Immerse the specimen in 30% (w/v) sucrose medium at 4 °C, until it sinks to the bottom of the container. Freeze as described in Method 2.
3. Cut sections as described in Method 3.
4. Let cryostat sections equilibrate and dry for 15 min at room temperature.
5. Preincubate sections for 30 min in a solution of 0.3% Triton X-100 for 30 min at room temperature.
6. Add 10 μl of NADPH solution (final concentration 1 mM) and 10 μl of NBT solution (final concentration 0.3 mM) to 1 ml of the Triton X-100 solution.
7. Incubate sections for 90 min at room temperature.
8. Rinse thoroughly with 0.1 M phosphate buffer (pH 5.3).
9. Mount in glycerol/gelatin.

Control incubations should be carried out (i) in the absence of substrate; (ii) in the presence of 5 mM dicumarol; (iii) in the presence of both 5 mM $NADP^+$ and 1 mM NADPH (should have no effect).

Method 7. Tetrazolium salt method for the demonstration of xanthine oxidoreductase (EC 1.1.1.204) activity

Materials

- Polyvinyl alcohol
- 0.1 M Phosphate buffer (pH 8.0)
- Hypoxanthine solution (6.8 mg/ml in 0.25 M NaOH)

- Methoxyphenazine methosulphate solution (15 mg/ml in distilled water)
- TNBT or NBT solution
  - Dissolve 5 mg in 20 µl of ethanol and 20 µl of dimethylformamide by gentle heating.
- 0.1 M Phosphate buffer (pH 5.3)
- Glycerol/gelatin solution [50% (v/v) glycerol/7% (w/v) gelatin]
- Allopurinol solution (13.6 mg/ml in 0.25 M NaOH)

Procedure

1. Use unfixed cell preparations or cryostat sections.
2. Dissolve 18 g of polyvinyl alcohol in 0.1 M phosphate buffer (pH 8.0) according to Method 4.
3. Add 10 µl of hypoxanthine solution (final concentration 0.5 mM), 10 µl of methoxyphenazine methosulphate solution (final concentration 0.45 mM) and 40 µl of TNBT or NBT solution (final concentration 5 mM) to 1 ml of polyvinyl alcohol-containing buffer.
4. Incubate fresh sections in the incubation medium for 30 min at 37 °C immediately after cutting.
5. Rinse thoroughly with 0.1 M phosphate buffer (pH 5.3) at 60 °C.
6. Mount in glycerol/gelatin.
7. Control incubations should be carried out (i) in the absence of substrate; and (ii) in the presence of 1 mM allopurinol.

Method 8. Demonstration of peroxisomal oxidase activity

Materials

- 0.1 M Tris/maleate buffer (pH 7.8)
- Cerium chloride solution (371 mg/ml in distilled water)
- Sodium azide solution (325 mg/ml in distilled water)
- Substrate solution (see Table 2)
- Distilled water
- 3,3′-Diaminobenzidine tetrahydrochloride

- 0.05 M Sodium acetate buffer (pH 5.3)
- Cobalt chloride
- 30% Hydrogen peroxide
- Glycerol/gelatin solution [50% (v/v) glycerol/7% (w/v) gelatin]

Procedure

1. Use unfixed cell preparations or cryostat sections. Let them equilibrate and dry for 5 min at 37 °C.

2. Prepare incubation medium by heating the buffer to 70 °C and adding 10 μl of cerium chloride solution (also heated to 70 °C; final concentration 10 mM) to 1 ml of Tris/maleate buffer. Cool to 37 °C and add 20 μl of sodium azide solution (final concentration 100 mM) and 20 μl of substrate solution.

3. Incubate for 30–120 min at 37 °C.

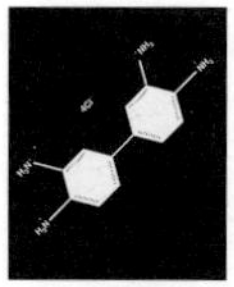

4. Rinse three times with distilled water.

5. Prepare the visualization medium by first dissolving of 50 mg of 3,3′-diaminobenzidine tetrahydrochloride in 100 ml of the sodium acetate buffer (final concentration 1.4 mM). Add 2 ml of sodium azide solution (final concentration 100 mM), 1 g of cobalt chloride (final concentration 42 mM) and, immediately before use, 7 μl of 30% hydrogen peroxide (final concentration 0.6 mM) to 100 ml of this buffer.

6. Incubate the cell preparations or cryostat sections for 30 min at room temperature in the visualization medium.

7. Rinse in distilled water.

8. Mount in glycerol/gelatin.

9. Store sections in the dark at 4 °C.

10. Control incubations should be carried out (i) in the absence of substrate; and (ii) in the presence of a specific inhibitor of the oxidase (see Table 2).

Method 9. Diaminobenzidine method for the demonstration of catalase (EC 1.11.1.6) activity

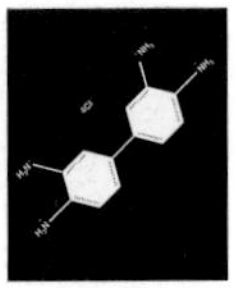

Materials

- 0.3% (w/v) Glutaraldehyde in distilled water
- Distilled water
- Polyvinyl alcohol
- 0.1 M Glycine/NaOH buffer (pH 10.5)
- 3,3′-Diaminobenzidine tetrahydrochloride
- 30% Hydrogen peroxide
- Glycerol/gelatin solution [50% (v/v) glycerol/7% (w/v) gelatin]
- 3-Amino-1,2,4-triazole solution (420 mg/ml in distilled water)

Procedure

1. Use unfixed cell preparations or cryostat sections. Let them equilibrate and dry for 5 min at 37 °C.
2. Fix sections in 0.3% (w/v) glutaraldehyde for 5 min at room temperature.
3. Rinse three times in distilled water.
4. Prepare incubation medium by dissolving 2 g of polyvinyl alcohol in 100 ml of 0.1 M glycine/NaOH buffer (pH 10.5) according to Method 4. Immediately before use, add 1.8 mg 3,3′-diaminobenzidine tetrahydrochloride (final concentration 5 mM) and 2 μl of 30% hydrogen peroxide (final concentration 18 mM) to 1 ml of this buffer.
5. Incubate sections for 30 min at 37 °C.
6. Rinse thoroughly in distilled water.
7. Mount in glycerol/gelatin.
8. Control incubations should be carried out (i) in the absence of substrate (hydrogen peroxide); and (ii) after preincubation in 50 mM 3-amino-1,2,4-triazole and 18 mM hydrogen peroxide.

Table 2. Recommended concentrations of substrates and inhibitors for histochemical incubation media to demonstrate activity of oxidases with the cerium method

Enzyme	Substrate	Inhibitor
α-Hydroxy acid oxidase (EC 1.1.3.15)	20 mM D,L-α-Hydroxybutyric acid or 1 mM glycolic acid	20 mM Phenylacetic acid
Xanthine oxidase (EC 1.1.3.22)	1 mM Hypoxanthine	1 mM Allopurinol
D-Amino acid oxidase (EC 1.4.3.3)	20 mM D-Proline	20 mM D,L-β-Hydroxybutyrate or 20 mM propargylglycine
Monoamine oxidase (EC 1.4.3.4)	6.25 mM Tryptamine/HCl	6.25 mM Iproniazid phosphate or pargyline
Urate oxidase (EC 1.7.3.3)	0.1 mM Uric acid	0.1 mM 2,6,8-Trichloropurine or 0.1 mM oxonic acid
Polyamine oxidase (EC number not yet assigned)	5 mM *N*′-Acetylspermidine, 5 mM *N*′-acetylspermine	5 mM 1,4-Butanediamine-*N*,*N*′-bis(2,3-butanedienyl)

Method 10. Diaminobenzidine method for the demonstration of peroxidase (EC 1.11.1.7) activity

Materials

- Polyvinyl alcohol
- 0.1 M Cacodylate buffer (pH 6.5)
- 3,3′-Diaminobenzidine tetrahydrochloride
- 30% Hydrogen peroxide
- Glycerol/gelatin solution [50% (v/v) glycerol/7% (w/v) gelatin]

Procedure

1. Use unfixed cell preparations or cryostat sections. Let them equilibrate and dry for 5 min at room temperature.

2. Prepare the incubation medium by dissolving 2 g of polyvinyl alcohol in 100 ml of 0.1 M cacodylate buffer (pH 6.5) according to Method 4. Immediately before use, add 90 mg of 3,3′-diaminobenzidine tetrahydrochloride (final concentration 2.5 mM) and 67 μl of 30% hydrogen peroxide (final concentration 6 mM).

3. Incubate the cell preparations or cryostat sections for 5 min at room temperature.

4. Rinse thoroughly in distilled water.
5. Mount in glycerol/gelatin.
6. Control incubations should be carried out (i) in the absence of substrate (hydrogen peroxide); and (ii) after preincubation in 18 mM hydrogen peroxide.

4.2. Transferases

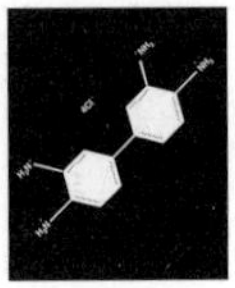

Transferases are a large group of enzymes that transfer a molecule from one substrate to another. Methods are available to localize histochemically the activities of specific transferases. Here, we present a method to localize kinase activities (Method 11 and Table 3). A kinase transfers a phosphate group from ATP or ADP to its substrate. Kinases are found in many metabolic pathways. Kinases involved in signal transduction pathways are of particular interest.

In Method 11, an auxiliary enzyme is present in the form of a thin film on the glass slide. When an auxiliary enzyme is used in this way, only a superficial layer of the section adjacent to the film of the auxilliary enzyme is involved in the histochemical reaction, owing to the limited penetration of the enzyme from the film into the tissue.

Method 11. Tetrazolium salt method for the demonstration of kinase activity using an auxiliary enzyme(s)

Materials

- Lyophilized pure auxiliary enzyme(s) (see Table 3)
- Distilled water
- Bovine serum albumin
- Brush
- Glass slides
- Polyvinyl alcohol

- 100 mM Tris/maleate buffer (pH 7–8)
- Sodium azide solution (33 mg/ml in distilled water)
- 1-Methoxyphenazine methosulphate solution (15 mg/ml in distilled water)
- Phenazine methosulphate solution (8 mg/ml in distilled water)
- TNBT or NBT solution
  - Dissolve 5 mg in 20 μl of ethanol and 20 μl dimethylformamide by gentle heating.
- 0.1 M Phosphate buffer (pH 5.3)
- Glycerol/gelatin solution [50% (v/v) glycerol/7% (w/v) gelatin]
- Stock solutions of substrate, co-enzyme and other constituents (see Table 3 for specific compounds and concentrations)

Procedure

1. Dissolve lyophilized pure auxiliary enzyme(s) in distilled water and add bovine serum albumin [final concentration 0.1% (w/v)].
2. Spread 10 μl of the enzyme solution with a brush to prepare a film on glass slides over an area of approximately 1 cm^2.
3. Dry the enzyme films in air and, when dry, use the slides immediately.
4. Pick up cryostat sections on the enzyme films on the glass slides.
5. Let the cryostat sections equilibrate and dry for 5 min at 37 °C.
6. Prepare the incubation medium by dissolving 18 g of polyvinyl alcohol in 100 mM Tris/maleate buffer (pH 7–8) according to Method 4. To 1 ml of this medium, add 10 μl of 100× concentrated stock solutions of substrate, co-enzyme, other constituents (see Table 3), sodium azide (final concentration 5 mM), either 1-methoxyphenazine methosulphate (final concentration 0.45 mM) or phenazine methosulphate (final concentration 0.2 mM) and 40 μl of TNBT or NBT (final concentration 5 mM).
7. Incubate sections for 5–30 min at 37 °C.
8. Rinse thoroughly with 0.1 M phosphate buffer (pH 5.3) at 60 °C.
9. Mount in glycerol/gelatin.

10. Control incubations should be carried out (i) in the absence of substrate(s); and (ii) in the presence of inhibitor(s) (see Table 3).

4.3. Hydrolases

Hydrolases are enzymes that degrade their substrates. Phosphatases split off phosphate groups from their substrate and are thus the counterpart of kinases. The activity of phosphatases is best demonstrated by Method 12 using the conditions listed in Table 4.

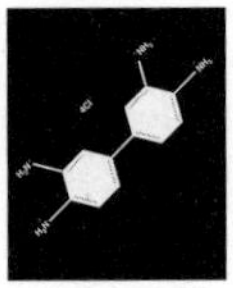

Glycosidases break down sugars, and Method 13 and Table 5 describe how their activity can be localized. This method can also be used to demonstrate the activities of esterases, enzymes that hydrolyse ester bonds. The activity of a specific esterase, acetylcholinesterase, which is present in synapses, can be demonstrated with Method 14.

Finally, protease activity is demonstrated using either a fluorescence technique, as described in Method 15 and Table 6, or with a diazonium salt method, as described in Method 16 and Table 6. Proteases are a large family of enzymes that can hydrolyse peptide bonds. These enzymes are involved in many physiological and pathophysiological processes such as activation of proteins by cleaving an amino acid sequence, processing of antigens for antigen presentation and catabolism of proteins, for example in the lysosomes. Proteases are usually synthesized in an inactive pro-form. Post-translational activation is mediated either by other proteases or by autocatalytic activity. Fig. 1 gives an example of post-translational activation of a proteinase. Fig. 1(A) shows the total amount of the inactive and active forms of a proteinase, cathepsin B, as demonstrated immunohistochemically, whereas Fig. 1(B) shows the metabolically active enzyme only. The method used for localization of cathepsin B activity as shown in Fig. 1(B) is presented in Method 15.

Table 3. Recommended concentrations of substrates, auxiliary enzymes and inhibitors for histochemical incubation media to demonstrate activity of kinases with the tetrazolium salt method

Enzyme	Substrates	Auxiliary enzymes	Inhibitors
Hexokinase (EC 2.7.1.1)	5 mM Glucose, 7.5 mM ATP, 1.5 mM NAD^+	Glucose-6-phosphate dehydrogenase	5 mM *N*-Acetylglucosamine or lauric acid
Phosphofructokinase (EC 2.7.1.11)	20 mM Fructose-6-phosphate, 2 mM ATP, 6 mM NAD^+	Fructose biphosphate aldolase, triose phosphate isomerase, glyceraldehyde-3-phosphate dehydrogenase	Phosphoenolpyruvate, arginine phosphate or Ca^{2+} (20 mM each)
NAD^+-kinase (EC 2.7.1.23)	1.5 mM NAD^+, 2.5 mM ATP	Glucose-6-phosphate dehydrogenase	Iodoacetamide, iodoacetate, ADP–ribose, ADP or AMP (2.5 mM each)
Pyruvate kinase (EC 2.7.1.40)	25 mM Phosphoenolpyruvate, 2.5 mM ADP, 1.9 mM NAD^+	Hexokinase, glucose-6-phosphate dehydrogenase	Glutathione, pyruvic acid or quercetin (25 mM each)
Creatine kinase (EC 2.7.3.2)	20 mM Phosphocreatine, 1.0 mM ADP, 0.45 mM NAD^+	Hexokinase, glucose-6-phosphate dehydrogenase	Dinitrofluorobenzene, sodium acetate or adenosine (20 mM each)

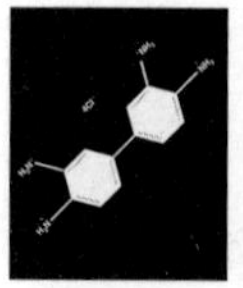

Table 4. Recommended substrate and inhibitor concentrations, and pH of histochemical incubation media, for demonstrating phosphatase activity with the cerium salt method

Enzyme	Substrate	Inhibitor	pH
Alkaline phosphatase (EC 3.1.3.1)	5 mM *p*-Nitrophenyl phosphate	5 mM Tetramisole	9.0
Acid phosphatase (EC 3.1.3.2)	5 mM *p*-Nitrophenyl phosphate or 5 mM β-glycerophosphate	Sodium fluoride or sodium molybdate or L-tartrate (5 mM each)	5.0
5′-Nucleotidase (EC 3.1.3.5)	5 mM AMP	Sodium fluoride or EDTA (5 mM each)	7.2
ADPase (EC no. not yet assigned)	5 mM ADP	–	7.2
Glucose-6-phosphatase (EC 3.1.3.9)	5 mM Glucose 6-phosphate	Citric acid (5 mM)	6.5
ATPases (EC 3.6.1.3)	5 mM ATP	Ouabain (5 mM) for Na^+,K^+-ATPase; EDTA (5 mM) for Ca^{2+}-ATPase	9.0
Thiamine pyrophosphatase (EC 3.6.1.–)	5 mM Thiamine pyrophosphate	–	7.4
Adenylate cyclase (EC 4.6.1.1)	1 mM AMP–P(CH_2)P	Alloxan (1 mM)	7.2

For specificity of the phosphatase reactions, series of control incubations with various inhibitors and activators have to be performed as described by Van Noorden and Frederiks [4].

Method 12. Cerium salt method for the demonstration of phosphatase activity

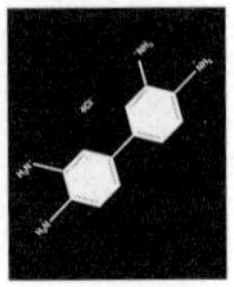

Materials

- 0.1 M Tris/maleate buffer (for pH, see Table 4)
- Cerium chloride
- Substrate (see Table 4)
- 0.05 M Tris/maleate buffer (pH 8.0)
- 3,3′-Diaminobenzidine tetrahydrochloride
- Sodium azide
- 30% Hydrogen peroxide
- Glycerol/gelatin solution [50% (v/v) glycerol/7% (w/v) gelatin in distilled water]
- Inhibitor of the phosphatase (see Table 4)

Procedure

1. Use unfixed cell preparations or cryostat sections. Let them equilibrate and dry for 5 min at 37 °C.

2. Prepare the incubation medium by dissolving 370 mg of cerium chloride (final concentration 10 mM) and substrate (see Table 4) in 100 ml of 0.1 M Tris/maleate buffer (for pH, see Table 4).

3. Incubate the cell preparations or cryostat sections for 5–60 min at 37 °C.

4. Rinse in 0.05 M Tris/maleate buffer (pH 8.0).

5. Prepare the second step medium by dissolving 360 mg of 3,3′-diaminobenzidine tetrahydrochloride (final concentration 10 mM) and 65 mg of sodium azide (final concentration 10 mM) in 100 ml of 0.05 M Tris/maleate buffer (pH 8.0). Immediately before use, add 10 ml of 30% hydrogen peroxide to the mixture (final concentration 1 M).

6. Incubate the cell preparations or cryostat sections for 10 min at room temperature in the second step medium.

7. Rinse in distilled water.

8. Mount in glycerol/gelatin.

9. Control incubations should be carried out (i) in the absence of substrate; and (ii) in the presence of a specific inhibitor of the phosphatase (Table 4).

Table 5. Recommended substrates and inhibitors, and pH of histochemical incubation media for the demonstration of glycosidase and esterase activities by the diazonium salt method

Enzyme	Substrate	Inhibitor	pH
Acid α-glucosidase (EC 3.2.1.20)	2-Naphthyl α-glucoside	Sucrose	6.0
α-Galactosidase (EC 3.2.1.22)	1-Naphthyl α-galactoside	Galactose	5.0
β-Galactosidase (EC 3.2.1.23)	1-Naphthyl β-galactoside	D-Galactono-1,4-lactone, *p*-chloromercuribenzoic acid or $HgCl_2$	4.0
Lactase (EC 3.2.1.23)	1-Naphthyl β-glucoside	D-Gluconolactone	6.0
α-Mannosidase (EC 3.2.1.24)	1-Naphthyl α-mannoside	Mannonolactone	4.5
N-Acetyl-β-glucosaminidase (EC 3.2.1.52)	Naphthol AS-BI-*N*-acetyl-β-glucosaminide	2-Acetamido-2-deoxy-D-gluconolactone	4.5
β-Glucuronidase (EC 3.2.1.31)	Naphthol AS-BI-glucuronide	D-Glucuronolactone, D-saccharonolactone, citrate or ascorbic acid	5.0
Non-specific esterase (EC 3.1.1.-)	α-Naphthyl acetate, α-naphthyl butyrate, naphthol AS-D-acetate, naphthol AS-D-chloroacetate	Physostigmine, di-isopropyl fluorophosphate or diethyl-*p*-nitrophenyl phosphate	5–8

Substrates and inhibitors are used at a final concentration of 5 mM.

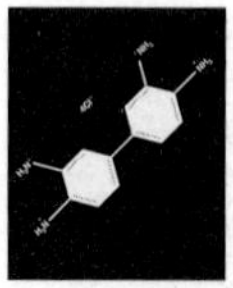

Method 13. Diazonium salt method for the demonstration of glycosidase and esterase activity

Materials

- Hexazotized pararosaniline
  - Basic fuchsine (CI 42500)
  - 2 M HCl
- Sodium nitrite solution (80 mg/ml in distilled water)
- Polyvinyl alcohol
- 0.1 M Sodium acetate buffer (for pH, see Table 5)
- Substrate solution (see Table 5)
- Dimethylformamide

- Distilled water
- Glycerol/gelatin solution [50% (v/v) glycerol/7% (w/v) gelatin]
- Specific inhibitor of the glycosidase or esterase (see Table 5)

Procedure

1. Use unfixed cell preparations or cryostat sections. Let them equilibrate and dry for 5 min at 37 °C.
2. Prepare hexazotized pararosaniline by dissolving 800 mg of basic fuchsine in 1 ml of 2 M HCl. Mix thoroughly until dissolved (for 2–3 h) and filter. The solution is stable for several months when stored at 4 °C in the dark. Filter again before use. As soon as possible before incubation, mix 1 ml of the solution with 1 ml of an aqueous solution of sodium nitrite.

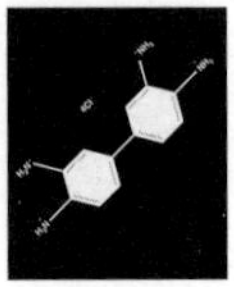

3. Prepare the incubation medium by dissolving 18 g of polyvinyl alcohol in 100 ml of 0.1 M sodium acetate buffer (for pH, see Table 5), according to Method 4. To 1 ml of this polyvinyl alcohol-containing buffer, add the substrate dissolved in 100 μl of dimethylformamide (see Table 5; final concentration 5 mM) and 60 μl of hexazotized pararosaniline (final concentration 10 mM).
4. Incubate the cell preparations or cryostat sections for 30–60 min at 37 °C.
5. Rinse with distilled water at 60 °C.
6. Rinse with distilled water.
7. Mount in glycerol/gelatin.
8. Control incubations should be carried out in the presence of a specific inhibitor of the glycosidase or esterase (see Table 5).

Method 14. Thiocholine method for the demonstration of acetylcholinesterase (EC 3.1.1.7) activity

Materials

- Polyvinyl alcohol
- 0.2 M Hepes buffer (pH 7.0)
- Sodium citrate solution (882 mg/ml in distilled water)
- Cupric sulphate solution (450 mg/ml in distilled water)
- Potassium ferricyanide solution (127 mg/ml in distilled water)
- Acetylthiocholine iodide solution (88 mg/ml in distilled water)
- Distilled water
- Glycerol/gelatin solution [50% (v/v) glycerol/7% (w/v) gelatin]
- Inhibitor

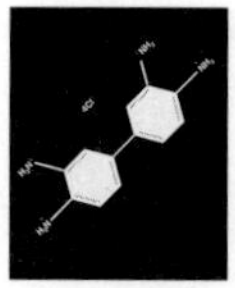

Procedure

1. Use unfixed cell preparations or cryostat sections. Let them equilibrate and dry for 5 min at 37 °C.
2. Dissolve 18 g of polyvinyl alcohol in 100 ml of Hepes buffer (pH 7.0) according to Method 4.
3. To 1 ml of polyvinyl alcohol-containing medium, add (strictly in the following order): 10 μl of sodium citrate (final concentration 30 mM), 10 μl of cupric sulphate (final concentration 18 mM), 10 μl of potassium ferricyanide (final concentration 3 mM), and 10 μl of acetylthiocholine iodide (final concentration 3 mM).
4. Adjust the pH of the medium to pH 6.0.
5. Incubate the cell preparations or cryostat sections in this medium for 60 min at 37 °C.
6. Rinse thoroughly with distilled water at 60 °C.
7. Mount in glycerol/gelatin.

8. Control incubations should be carried out (i) in the absence of substrate; and (ii) in the presence of 3 mM of an inhibitor such as physostigmine, di-isopropyl fluorophosphate, tetraisopropyl phosphoramide or 1,5-bis-(4-trimethylammoniumphenyl)-pentane-3-on-di-iodide.

Method 15. Fluorescence method using 5′-nitrosalicylaldehyde for the demonstration of protease activity

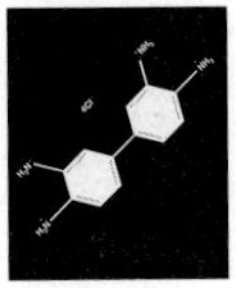

Materials

- 2-Hydroxy-5-nitrobenzaldehyde (5′-nitrosalicylaldehyde; Merck, Darmstadt, Germany) solution (17 mg in 200 μl of diethylether and 800 μl of dimethylformamide)
- Diethylether
- Dimethylformamide
- Substrate solution (1 mg dissolved in 20 μl of dimethylformamide)(see Table 6 for specific substrates; Enzymes Systems Products, Livermore, CA, U.S.A. and Bachem, Bubendorf, Switzerland)
- 100 mM Tris/HCl buffer (pH 7.2)
- 100 mM Sodium cacodylate buffer (for pH see Table 6)
- NaCl
- 100 mM Phosphate buffer (pH 6.0)
- Dithiothreitol solution (15 mg/ml in distilled water)
- EDTA disodium salt solution (50 mg/ml in distilled water)
- L-Cysteine solution (32 mg/ml in distilled water)
- Inhibitor solution [1 mg dissolved in 20 μl of dimethylformamide (see Table 6 for specific inhibitors; Enzymes Systems Products and Bachem)]
- Fluorescence microscope with an excitation filter of 440–500 nm, a barrier filter of 515 nm and a water-immersion objective (×40)

Procedure

1. Use unfixed cell preparations or cryostat sections. Let them equilibrate and dry for 10 min at room temperature.

2. Preparation of incubation media.

(a) *γ-Glutamyltransferase, aminopeptidase or dipeptidylpeptidase activity*: Add 10 μl of 2-hydroxy-5-nitrobenzaldehyde solution (final concentration 1 mM) and 1 mg of the substrate (see Table 6) dissolved in 20 μl of dimethylformamide (final concentration 2–3 mM) to 1 ml of 100 mM sodium cacodylate buffer (for pH, see Table 6).

(b) *Elastase activity*: Add 30.5 mg of NaCl (final concentration 0.5 M), 10 μl of 2-hydroxy-5-nitrobenzaldehyde solution (final concentration 1 mM), and Suc-Ala-Ala-Ala-4-methoxy-2-naphthylamide (final concentration 2 mM) to 1 ml of 100 mM Tris/HCl buffer (pH 7.2).

(c) *Cathepsin B activity*: Add 10 μl of dithiothreitol (final concentration 1 mM), 10 μl of EDTA disodium salt (final concentration 1.3 mM), 10 μl of L-cysteine (final concentration 2.7 mM), 10 μl of 2-hydroxy-5-nitrobenzaldehyde (final concentration 1 mM) and *N*-Cbz-Ala-Arg-Arg-4-methoxy-2-naphthylamide (final concentration 2 mM) to 1 mM phosphate buffer (pH 6.0).

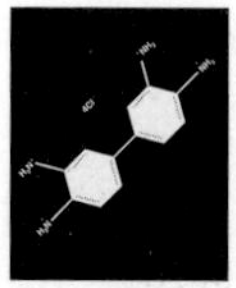

3. Pour a drop of incubation medium on to the cell preparation or section (50–100 μl) to start the reaction. Monitor the incubation time.

4. Put the slide with the cells or the tissue section immersed with incubation medium onto the stage the fluorescence microscope. Use a water-immersion objective, preferably 40×. Use the incubation medium as the drop of 'water' for the water-immersion objective.

5. Monitor development of yellow fluorescence as a product of protease activity by incubating at room temperature on the stage of a fluorescence (photo)microscope at regular intervals (1–5 min) for up to 120 min. Yellow fluorescence usually starts to appear after 10 min. In most tissues, 15–20 min of incubation is best for analysis of protease activity.

   Green fluorescence is due to non-specific binding of 5′-nitrosalicylaldehyde to proteins.

Monitoring of the formation of the yellow fluorescent reaction product is necessary because this product diffuses and recrystallizes rapidly during rinsing and mounting of the preparations.

6. Take photomicrographs or capture digital images during incubation.
7. Control incubations should be carried out in the presence of specific inhibitors (see Table 6).

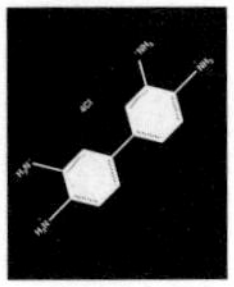

Table 6. Recommended substrates [with 4-methoxy-2-naphthyl amide (MNA) as the leaving group], inhibitors and pH of histochemical incubation media for the demonstration of protease activity

Enzyme	Substrate	Inhibitor	pH
γ-Glutamyltransferase (EC 2.3.2.2)	γ-Glutamyl-MNA	*p*-Chloromercuribenzoic acid, iodoacetic acid or magnesium ions	8.0
Aminopeptidase M (EC 3.4.11.2)	Leucyl-MNA	Puromycin, EDTA or 1,10-phenanthroline	8.0
Aminopeptidase A (EC 3.4.11.7)	α-Glutamyl-MNA	Angiotensin	7.0
Dipeptidylpeptidase II (EC 3.4.14.2)	Lysyl-alanyl-MNA	Puromycin, Tris	6.0
Dipeptidylpeptidase IV (EC 3.4.14.5)	Glycyl-prolyl-MNA	Di-isopropyl fluorophosphate, diethyl-*p*-nitrophenyl phosphate (E600) or Ala–Pip^{P}(OPh-4-Cl)$_2$	8.0
Elastase (EC 3.4.21.36/37)	Suc-alanyl-alanyl-alanyl-MNA	Phosphate, antileukoprotease or L658.758	7.0
Cathepsin B (EC 3.4.22.1)	*N*-Cbz-alanyl-arginyl-arginyl-MNA	Leupeptin, *trans*-epoxysuccinyl-L-leucyl-amido-(4-guanidino)-butane (E64) or phenylhomophenyl-fluoromethyl ketone	6.0
Tryptase (EC 3.4.21.59)	D-Valyl-leucyl-arginyl-MNA	*p*-Aminobenzamidine or 4-amidino-phenylmethane-sulphonyl fluoride	7.4
Urokinase-type plasminogen activator (EC number 3.4.21.73)	Z-Glycyl-glycyl-arginyl-MNA	Amiloride	7.4

The substrate concentration for each method is 1 mg/ml of incubation medium (final concentration 2–3 mM). Inhibitors, when used, should also be used at 2–3 mM. Cbz, benzyloxycarboxyl; Suc, succinyl.

Method 16. Diazonium salt method for the demonstration of protease activity

Materials

- Polyvinyl alcohol
- 0.1 M Sodium cacodylate buffer (for pH, see Table 6)

- Dimethylformamide
- Substrate solution [1 mg dissolved in 20 μl of dimethylformamide; see Table 6 for specific substrates (Enzymes Systems Products and Bachem)]
- Fast Blue BB salt (CI 37175; Serva, Heidelberg, Germany)
- Distilled water
- Cupric sulphate solution (25 mg in 1 ml of distilled water)
- Glycerol/gelatin solution [50% (v/v) glycerol/7% (w/v) gelatin]
- Inhibitor solution [1 mg dissolved in 20 μl of dimethylformamide; see Table 6 for specific inhibitors (Enzymes Systems Products and Bachem)]

Procedure

1. Use unfixed cell preparations or cryostat sections. Let them equilibrate and dry for 5 min at 37 °C.

2. Prepare the incubation medium by dissolving 18 g of polyvinyl alcohol in 100 ml of 0.1 M sodium cacodylate buffer (as described in Method 4). Add 1 mg of substrate dissolved in 100 μl of dimethylformamide (see Table 6; final concentration 2–3 mM) and 5 mg of Fast Blue BB salt dissolved in 60 μl of dimethylformamide (final concentration 12.5 mM).

3. Incubate the cell preparations or cryostat sections for 10 min at 37 °C.

4. Rinse with distilled water at 60 °C.

5. Chelate the final reaction product with cupric sulphate (final concentration 100 mM) for 5 min at room temperature to further insolubilize the final reaction product and to minimize recrystallization.

6. Rinse with distilled water.

7. Mount in glycerol/gelatin.

8. Control incubations should be carried out in the presence of specific inhibitors (see Table 6).

5. Troubleshooting

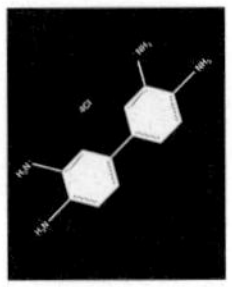

Problem	Diagnosis
No coloured reaction product is formed, or the coloured reaction product is localized diffusely as a smear, and not localized precisely.	1. One of the compounds of the incubation medium may have precipitated. Check whether the medium is completely clear. If not, prepare new medium and make sure all compounds remain in solution and that the medium is completely clear before use. 2. Compounds have not been thoroughly mixed in the viscous polyvinyl alcohol-containing incubation medium. Use a spatula to ensure proper mixing.
No coloured reaction product is formed in cells.	Cell membranes are not sufficiently permeable to the compounds in the incubation medium. Try a freeze–thaw step before incubation.
Coloured reaction product is formed after the reaction has been stopped, and cells or sections have been mounted. This leads to a slow darkening of the preparation.	Rinsing off of the incubation medium was not performed thoroughly enough. Try to rinse as thoroughly as possible. Use warm rinsing media with a pH different from the pH that is optimum for the enzyme.
Coloured reaction product fades during storage.	Store the slides in a dark, cool (4–10 °C) place after staining and mounting.
Cells or sections are lost from the slide during incubation.	Cell preparation or section have not been dried for long enough before incubation. Prolong period of drying. If that does not help, use slides that are coated with poly-L-lysine or gelatin [4].
Coloured reaction product is formed in unexpected sites in cells or sections.	Check the specificity of the reaction by adding inhibitors of the enzyme under study to the incubation medium. Never forget to perform control reactions!

References

1. Van Noorden, C.J.F., Boonacker, E., Bissell, E.R., Meijer, A.J., Van Marle, J. and Smith, R.E. (1997) Anal. Biochem. **252**, 71–77

2. Boonacker, E.P. and Van Noorden, C.J.F. (2001) J. Histochem. Cytochem. **49**, 1473–1486

3. Jonker, A., Geerts, W.J.C., Charles, R., Lamers, W.H. and Van Noorden, C.J.F. (1995) J. Histochem. Cytochem. **43**, 1027–1034

4. Van Noorden, C.J.F. and Frederiks, W.M. (1992) Enzyme Histochemistry. A Laboratory Manual of Current Methods, Bios, Oxford

5. Stoward, P.J. and Pearse, A.G.E. (1991) Histochemistry, Theoretical and Applied. vol. 3, 4th edn, Churchill–Livingstone, Edinburgh

6. Chayen, J. and Bitensky, L. (1991) Practical Histochemistry, 2nd edn, Wiley, London

7. Lojda, Z., Gossrau, R. and Schiebler, T.H. (1979) Enzyme Histochemical Methods. A Laboratory Manual, Springer Verlag, New York

8. Lyon, H. (1991) Theory and Strategy in Histochemistry, Springer Verlag, Berlin

9. Hayhoe, F.G.J. and Quaglino, D. (1988) Haematological Cytochemistry, 2nd edn, Churchill–Livingstone, Edinburgh

10. Wohlrab, F. and Gossrau, R. (1992) Katalytische Enzymhistochemie. Grundlagen und Methoden für die Elektronenmikroskopie, Gustav Fischer Verlag, Jena

11. Van Noorden, C.J.F. and Hulstaert, C.E. (1991) in Electron Microscopy of Tissues, Cells, and Organelles (Harris, J.R., ed.), pp. 125–149, Oxford University Press, Oxford

12. Zollner, H. (1989) Handbook of Enzyme Inhibitors, V.C.H. Gesellschaft, Weinheim

13. Wehby, R.G. and Frank, M.E. (1999) J. Histochem. Cytochem. **47**, 197–207

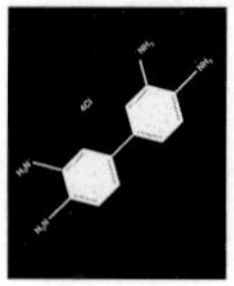

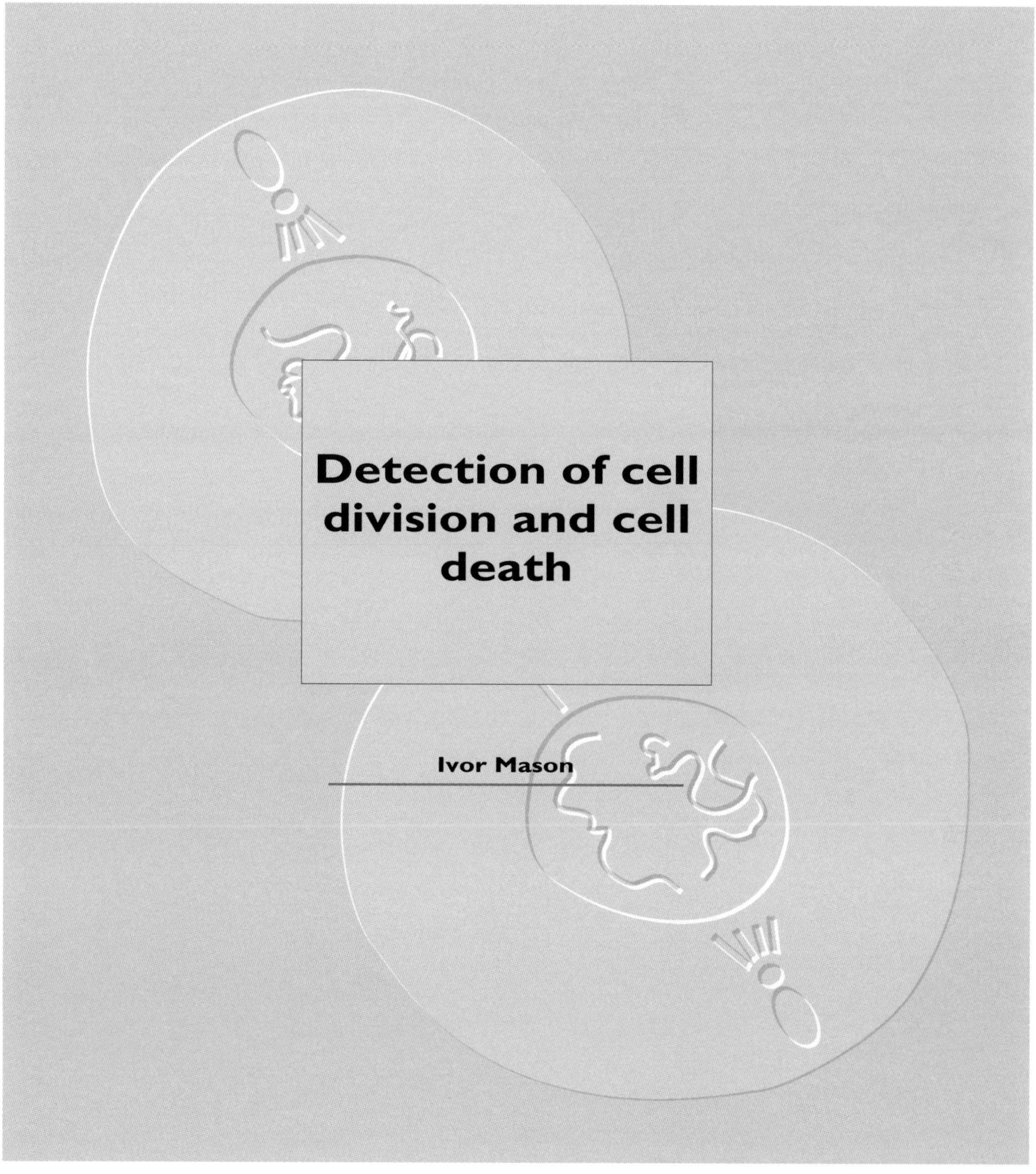

Detection of cell division and cell death

Ivor Mason

Chapter 12

1. Introduction

The final cell number in a tissue is determined by the balance of cell division and cell death. Measurement or detection of both are important to a number of aspects of biological research, including tumour growth and killing, growth and sculpting of the embryonic form, and immune functions. The following are a series of techniques that can be combined with microscopy to detect and/or quantify division or apoptosis.

2. Techniques for staining dividing cells

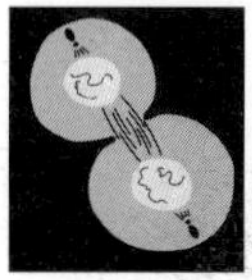

Dividing cells can be detected either by the incorporation of a labelled nucleotide {usually bromo-deoxyuridine (BrdU), although [^{3}H]thymidine was previously widely used} into nuclear DNA (Fig. 1A) or by detection of proteins whose expression or modification is associated with dividing cells. Such proteins or their modification are only associated with certain phases of the cell cycle. However, if a value for the mitotic index of a cell population is required then there is no alternative to labelling with a pulse of BrdU delivered over a defined period of time and then chased with unlabelled dUTP.

2.1 Incorporation of BrdU

Method 1. Labelling animals with BrdU

Materials

- Phosphate-buffered saline (PBS)
  - 10× PBS: 1.3 M NaCl, 70 mM Na_2HPO_4, 30 mM NaH_2PO_4. Adjust pH to 7.0 with dilute HCl or NaOH as appropriate.

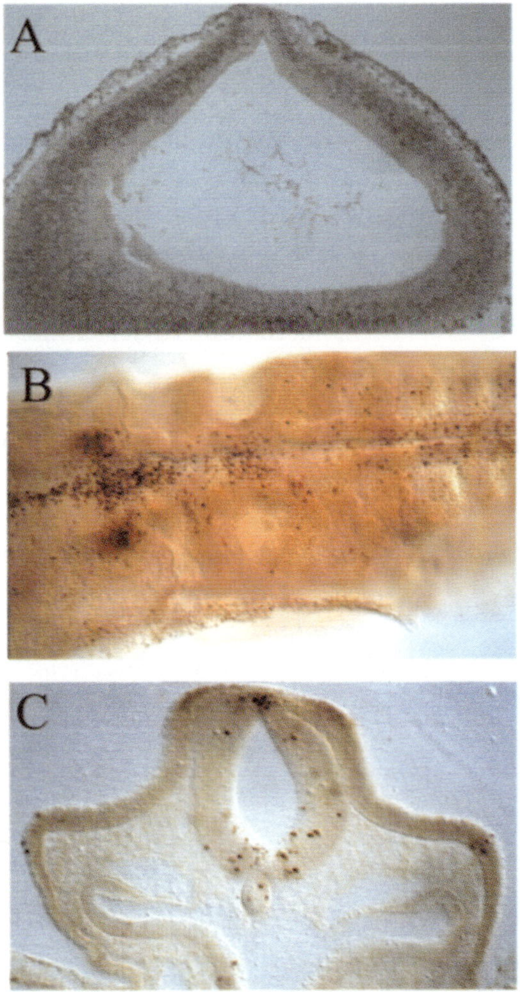

Fig. 1. Staining of dividing and dead or dying cells

(A) Section taken through the developing midbrain of a 3.5-day chick embryo showing incorporation of BrdU (dark blue nuclei) during a 1 h pulse. (B) TUNEL staining of the hindbrain region of a whole 2.5-day chick embryo showing prominent cell death (brown nuclei) in the region of the dorsal neural tube. (C) A transverse section taken through a 2.5-day chick embryo at the level of the otic pit and stained using the TUNEL process. Apoptotic nuclei (brown) are visible in the neural tube, notochord and lateral ectoderm.

- BrdU stock solution
  - 10 mM BrdU (Roche, Welwyn Garden City, U.K.) dissolved in PBS (pH 7.0).
- 4% (w/v) Paraformaldehyde (Fluka, Dorset, U.K.)
  - Add 16 g of paraformaldehyde (Fluka no. 76240) to 400 ml of 1× PBS (diluted in distilled water) and heat at 65 °C in a water bath with occasional swirling until dissolved (approx. 90 min). It is cooled to 4 °C prior to use or stored in 20ml aliquots at −20 °C.

> ***Caution** Paraformaldehyde must be weighed out wearing mask and gloves in a fume hood.*

Procedure

1. Inject the animal intraperitoneally with BrdU stock solution (1 ml per 100 g body weight).

   The worker must have appropriate licensing to perform such a manipulation and these regulations vary from country to country. This approach can also be used to label embryos following injection of pregnant females.

2. Kill the animal 2–4 h later (noting the time if mitotic index values are required).

3. Fix in 4% paraformaldehyde.

Method 2. Labelling tissue slices with BrdU

Materials

- Culture medium or Ringer's solution appropriate to the species being used
- CO_2 incubator
- BrdU stock solution (see Method 1)
- 4% (w/v) Paraformaldehyde (see Method 1)

Procedure

1. Dissect the tissue of choice in prewarmed medium to give slices approx. 1 mm thick.

2. Place these in prewarmed culture medium or Ringer containing 0.01 volumes of prewarmed BrdU stock solution.

3. Incubate at 37 °C (lower for amphibian or teleost tissue) in an incubator having a CO_2 atmosphere appropriate to the culture medium for 1 h.

4. Replace the labelling medium with fresh medium lacking BrdU and incubate for a further 30 min.

5. Fix in 4% paraformaldehyde and process for sectioning.

Method 3. Preparation of microscope slides

Materials

- Washed glass microscope slides and coverslips (Merck BDH, Poole, U.K.)
- 25-Place metal slide racks (R.A. Lamb, Eastbourne, U.K.)
- 400 ml capacity glass troughs with lids suitable for holding slide racks (R.A. Lamb)
- Concentrated HCl 10% (v/v) (Merck BDH)
- Acetone (AnalaR, Merck BDH)
- TESPA (3-aminopropyltriethoxysilane; Sigma-Aldrich, Poole, U.K.)

> ***Note:*** *Use TESPA in a fume hood.*

- 70% (v/v) Ethanol
- 95% (v/v) Ethanol
- Distilled water [this must not be treated with diethylpyrocarbonate (DEPC; Sigma-Aldrich)]
- Aluminium foil

Microscope slides need to be washed and silanized prior to collection of sections. Alternatively, pre-treated slides (e.g. Superfrost, Merck BDH) are now available commercially.

Procedure

1. Place metal racks of 25 slides in 70% ethanol, followed by concentrated HCl, distilled water, and finally 95% ethanol (400 ml of each) for 10 s each.
2. Wrap in aluminium foil to keep free from dust and place the slides at 150 °C for 30 min.
3. While still wrapped in foil, cool the slides on a bench.
4. Place the rack of slides, in acetone containing 2% (v/v) TESPA, then in two changes of acetone and finally in distilled water each for 10 s.

Step 4 should be performed in a fume cupboard.

5. Dry the foil-wrapped slides at 37 °C overnight.
6. Slides are stored individually in racks (to avoid scratches) at room temperature for up to 3 months.

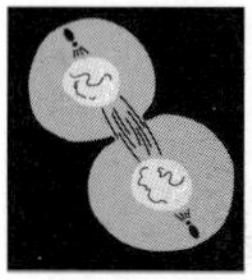

Method 4. Detection of BrdU in tissue sections

Materials

- 10× Saline
  - □ 8.3% (w/v) NaCl. Add DEPC to a concentration of 0.1%, shake well and autoclave.
- 100% Ethanol
- Graded ethanols [20%, 30%, 50%, 70%, 80% and 95% (v/v) ethanol] prepared with DEPC-treated water
- Xylene (AnalaR, Merck BDH)
- Histoclear (National Diagnostics, Hull, U.K.)
- Wax (e.g. Paramat pastillated wax, Merck BDH)
- Plastic moulds for wax
- Suitable rocking microtome with disposable blades (reduces chance of contamination with RNase from a generally-used blade)
- Bath for sections (well-cleaned with detergent, rinsed well with distilled water and filled with DEPC-treated water)

- Hot plate
- Trypsin solution
  - 0.05% (w/v) Trypsin (Sigma-Aldrich), 0.05% (w/v) $CaCl_2$ in 1× PBS
- 4 M HCl
- 0.5% (w/v) Bovine serum albumin (BSA; Roche), 0.1% (v/v) Tween-20 (Sigma-Aldrich) in PBS
- Anti-BrdU $F(ab)_2$ antibody fragments coupled to alkaline phosphatase (Roche)
- Nitro-Blue Tetrazolium (NBT) solution (Roche)
- 5-Bromo-4-chloro-indolyl phosphate (BCIP) solution (Roche)
- NTMT solution
  - 100 mM NaCl, 100 mM Tris/HCl (pH 9.5), 50 mM $MgCl_2$, 0.1% (w/v) Tween 20 (make fresh on day of use from concentrated stock of each individual component)
- Glycergel

Procedure

1. Fix labelled tissue overnight (from Methods 1 or 2).
2. Incubate in 1× PBS on ice for 30 min (small embryos: chicks <E6.5; mice <E13.5) or 1 h (larger material). Use glass containers as later treatments involve high temperatures and organic solvents.
3. Further dissect embryos or tissues if required, e.g. cut embryos away from surrounding membranes. Stage embryos if required.
4. Replace PBS with 1× saline, incubate for 30 min (small embryos/tissue pieces) or 1 h (larger material) on ice.
5. Replace saline with saline/ethanol (1:1, v/v) and incubate for 30 min (small embryos/tissue pieces) or 1 h (larger material) on ice.
6. Replace saline/ethanol with 70% ethanol and incubate for 30 min (small embryos) or 1 h (larger material) on ice.
7. Repeat step 6.
8. Replace 70% ethanol with 80% ethanol and incubate on ice for 30 min.

9. Replace with 95% ethanol and incubate for 30 min (small embryos/tissue pieces) or 1 h (larger material) on ice.
10. Place in 100% ethanol and incubate for 30 min (small embryos/tissue pieces) or 1 h (larger material) on ice.
11. Repeat Step 10
12. Replace with 100% ethanol for 30 min (small embryos/tissue pieces) or 2 h (larger material).
13. Replace 100% ethanol with xylene or (preferably) Histoclear in glass bottles and incubate on ice for 30 min (small embryos/tissue pieces) or 1 h (larger material).
14. Repeat step 13 with fresh xylene or Histoclear.
15. Pour off xylene or Histoclear and half-fill bottles with fresh molten wax (sections cut from blocks prepared with freshly melted wax pastules have better elasticity than from wax which has remained molten for days) and place at 52 °C for 45 min (small embryos) or 90 min (larger material).

16. Change wax twice; incubate each time for 60 min (small embryos/tissue pieces) or 90 min (larger material).
17. Embed the tissue. First, fill moulds with wax (if possible do this on a warmed plate to keep the wax molten). Carefully transfer the embryo to the mould, orient with a hot mounted needle or warm forceps, and move to a cooled plate or onto the bench. Label one end of the block for reference after the wax is set and the embryo's orientation is no longer visible (e.g. by inserting a piece of card on which details of the specimen and its orientation can be written) and allow the wax to set completely (1–2 h).

    It is crucial, during fixation, processing and embedding in wax prior to sectioning, that all of the reagents completely penetrate and equilibrate within the embryonic tissue. As mentioned earlier, times vary according to the size of the embryo/tissue being prepared. The following illustrates modifications that have been used successfully with larger tissues.

(i) Fix in 4% paraformaldhyde in 1× PBS overnight at 4 °C.

(ii) Wash twice in PBS for 30 min each and twice in saline for 30 min each. Incubate in 30% ethanol for 1 h, 50% ethanol for 1 h, 70% ethanol for 1 h, 100% ethanol for 3 h and then in fresh 100% ethanol overnight.

(iii) Incubate in fresh 100% ethanol for a further 5 h. Then incubate in three changes of Histoclear (or xylene): the first for 1 h, the second for 3 h and the third overnight.

(iv) Replace the Histoclear with freshly melted (52 °C) wax and incubate at 52 °C in an oven for 3 h. Replace with fresh wax and incubate at 52 °C for 5 h and then replace once more and incubate at 52 °C overnight.

(v) Incubate with two more changes of wax, each for 3 h, and then embed the tissue.

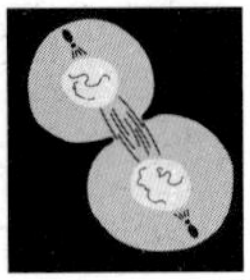

18. Cut 7–10 μm sections as ribbons on a rotary microtome using a disposable blade.

19. Ribbons are placed on a pool of 20% ethanol (in DEPC-treated water) on a microscope slide and cut into smaller strips with a clean razor blade if required.

20. The sections are floated out onto distilled water in a clean section bath at 45 °C until any creases disappear.

21. Sections are collected onto TESPA-treated slides (see Method 3), and dried at 48 °C on a hot plate for 2 h, or overnight at 37 °C.

    It is useful to collect at least two adjacent sections per slide; this aids the subsequent identification of stained tissues as they should give identical results on both sections.

22. Store desiccated slides at 4 °C so that they are free from moisture and that sections still give acceptable results months after being cut.

    Before use, allow slides to warm to room temperature.

23. Dewax slides in xylene twice for 10 min each in a fume cupboard.

24. Place in 100% ethanol for 2 min to remove most of the xylene.

25. Re-hydrate sections through the following graded alcohol series: 100% ethanol twice, 95% cthanol, 80% ethanol, 70% ethanol, 50% ethanol and 30% ethanol.

26. Place in 1× saline for 5 min and then in 1× PBS for 5 min.

27. Cover the section with trypsin solution and incubate for 10 min at 37 °C.

28. Wash twice for 5 min in PBS.

29. Cover specimen with 4 M HCl for 10–20 min.

30. Wash three times, for 5 min each, with PBS. Check that pH of last wash is above 6.5; if not, wash further.

31. Incubate in 0.5% (w/v) BSA/0.1% (v/v) Tween 20 in PBS for 15 min.

32. Cover the specimen with anti-BrdU antibody (50–200 μl) and dilute the antibody according to the manufacturer's recommendations in 0.5% (w/v) BSA/0.1% Tween 20 in PBS. Incubate for 2 h at room temperature.

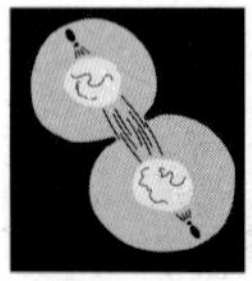

33. Wash three times in 0.5% (w/v) BSA/0.1% Tween 20 in PBS.

34. Wash three times for 15 min each with NTMT solution.

35. Incubate with NTMT containing 4.5 μl/ml NBT and 3.5 μl/ml BCIP stock solutions in the dark until desired reaction product intensity is achieved.

36. Stop the reaction by washing in PBS. Refix in 4% paraformaldehyde in 1× PBS for 1 h.

37. Mount under coverslips in glycergel.

2.2. Detection of proteins associated with cell division

A number of proteins are expressed in dividing cells but are absent from quiescent cells (in G_0 phase). Of these, some are only detectable during certain phases of the cell cycle (G_1, S, G_2 or M) and those expressed in M

Table 1. Antigens used to detect cell cycle phase

Antigen	Cell cycle phase					Supplier of antibody
	G_0	G_1	S	G_2	M	
Casein kinase 2	–	+	+	+	+	Roche
PCNA	–	+ [late G_1]	++	+	(+)	Roche
Ki-67	–	(+) [late G_1]	+	+	+	Roche
Topoisomerase II	–	++	+	(+)	(+)	Roche
Phosphorylation of histone H3	–	–	–	+	+	Upstate Biotechnology

PCNA, proliferating-cell nuclear antigen. (+), Weak staining; +, strong staining.

or G_2 phase are particularly useful in characterizing dividing cells in tissue sections. Table 1 details some of the more useful antigens.

Antibodies to many of these proteins are compatible with either cryostat or paraffin wax sections. However, this will vary from antibody to antibody, as will optimal fixation conditions and working concentration of antibody (see Chapters 6 and 9 in this volume). The supplier's datasheet should be consulted for the appropriate experimental procedure and species-specificity of the antibody.

3. Techniques for staining dead or dying cells

During the past decade there has been a resurgence of interest in cell death [1] — in particular in apoptosis (as opposed to necrosis) — as a mechanism of regulating cell numbers. Consequently, much of the intracellular machinery that regulates and mediates apoptosis has been identified, and there has been a near-exponential increase in publications on apoptosis. Approximately 40 proteins have been discovered whose known functions are exclusively to do with the initiation or regulation of apoptosis. In addition, a further 30 or so molecules, although already associated with important roles in signalling or DNA replication, transcription or repair, have also been recognized as regulating apoptosis. It is now clear that apoptosis plays as important a role in the regulation of final cell numbers as does cell division.

The simplest method to detect cell death is the use of a vital stain, e.g. Acridine Orange (see [2] for a protocol). However, these methods suffer the disadvantages that they are non-permanent, can only be performed on fresh (unfixed) tissue and do not distinguish between apoptosis and necrosis. The alternatives are to detect the fragmentation of chromosomal DNA that is characteristic of apoptosis, e.g. by terminal transferase-
... nick end labelling (TUNEL) (Figs. 1B and 1C) [3]
... sociated with that process.

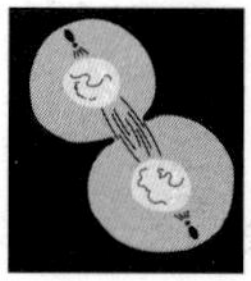

3.1 T...

... kits for performing TUNEL (Roche,
... s expensive alternative is for the researcher

Meth... e

... sition of 10× PBS.
... in 1× PBS (see Method 1)

> ***Note*** *Paraformaldehyde must be weighed out while wearing mask and gloves in a fume hood.*

- Proteinase K (Roche)
- 10 mM Tris/HCl (pH 7.4)
- Triton X-100 (Sigma-Aldrich)
- Terminal deoxynucleotide transferase (TdT) (Roche)
  - Supplied with 5× TdT buffer and 25 mM $CoCl_2$.
- 200 mM dUTP (Roche)
- 200 mM Biotin–16-deoxyuridine triphosphate (biotin–16-dUTP; Roche)
- Streptavidin–fluorescein conjugate (Amersham plc, Little Chalfont, U.K.)
- Streptavidin–horseradish peroxidase conjugate (Amersham plc)
- DABCO [1,4-diazobicyclo-(2,2,2)-octane] antiquenching mounting medium.
  - Dissolve 2.5 g of DABCO (Sigma-Aldrich) in 90 ml of glycerol at 37 °C and 0.1 g of sodium azide in 10 ml of 1× PBS. Combine the two solutions, mix well and store at 4 °C in the dark.
- 3′3′-Diaminobenzidine tetrahydrochloride (DAB) (Sigma-Aldrich)
- 30% (w/v) Hydrogen peroxide (Sigma-Aldrich)

Procedure

1. Fix embryos or tissue in 4% paraformaldehyde overnight at 4 °C.
2. Wash three times in PBS/1% (v/v) Triton X-100 for 30 min each (increase times for older embryos).
3. Incubate in 1× TdT buffer, 2.5 mM $CoCl_2$, 6 mM dUTP, 4 mM biotin–16-dUTP, 1% (v/v) Triton X-100 and 0.5 units/ml TdT.
4. Wash in PBS/1% (v/v) Triton X-100 three times for 30 min each (increase times for older embryos).
5. Incubate overnight with 1:100 dilution in PBS of streptavidin–fluorescein conjugate or streptavidin–horseradish peroxidase conjugate.
6. Wash in PBS/1% (v/v) Triton X-100 three times for 30 min each (increase times for older embryos).
7. (a) If using the fluorescein conjugate, equilibrate and mount the embryos in DABCO mountant and view using epifluorescence. (b) If using the

horseradish peroxidase conjugate, wash embryos twice for 30 min in PBS and then proceed as below.

8. Equilibrate in 5 mg/ml DAB in 1× PBS for 3 h by rocking in darkness.
9. Replace the above with 5 mg/ml DAB in PBS containing 2 μl/ml 30% (v/v) hydrogen peroxide and develop for 5–15 min in darkness.
10. When background staining begins to appear, stop the reaction with several changes of PBS.
11. Equilibrate and mount under a coverslip in 1× PBS/90% (v/v) glycerol.

Method 6. TUNEL for sections

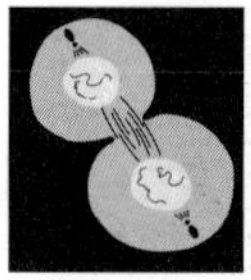

Materials

- 10× PBS
- 4% (w/v) Paraformaldehyde in PBS
- 10× Saline [8.3% (w/v) DEPC-treated NaCl]
- Howard's Ringer
  - 0.12 M NaCl
  - 0.0015 M $CaCl_2$
  - 0.005 M KCl
  - Adjust the pH to 7.2 with very dilute HCl.
- Absolute (100%) ethanol and graded ethanols [30%, 50%, 70%, 80% and 95% (v/v) prepared with DEPC-treated water]
- Xylene (AnalaR, Merck BDH)
- Histoclear (National Diagnostics)
- 10 mM Tris/HCl (pH 7.4)
- Proteinase K
- Wax (e.g. Paramat pastillated wax, Merck BDH)
- Plastic moulds for wax
- Suitable rocking microtome, disposable blades and holder, and bath for sections (well-cleaned with detergent, rinsed well with distilled water and filled with DEPC-treated water and hot plate)
- Triton X-100
- TdT (Roche)

- 5× TdT buffer (Roche)
- 25 mM $CoCl_2$
- 200 mM dUTP (Roche)
- 200 mM biotin–16-dUTP (Roche)
- Streptavidin–fluorescein conjugate (Amersham plc)
- Streptavidin–horseradish peroxidase conjugate (Amersham plc)
- DABCO (see Method 5)
- DAB (see Method 5)
- 30% (w/v) Hydrogen peroxide (Sigma-Aldrich)

Procedure

1. Dissect embryos or tissues in 1× PBS (mammals) or an appropriate Ringer's solution (other vertebrates). Fix in a large volume (at least five times the volume of tissue) of 4% (w/v) paraformaldehyde in PBS at 4°C overnight in a vessel which adequately contains the fumes.

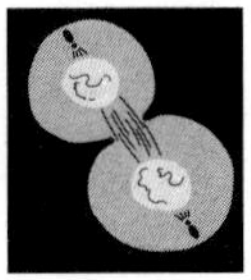

   Smaller embryos or tissue pieces can be fixed for shorter periods of time (4 h minimum) and it is convenient to dechorionate zebrafish after fixation. To prevent them from rolling up, embryos can be fixed pinned out in Slygard-coated Petri dishes if required in the designated fridge. Fixation times will vary with the size of the embryo or tissue from a few hours (mice embryos up to E9.5, chicks to stage 18) to 12 h (E18 mice embryos) or even longer (newborn mice or chicks from stage 38). In our experience, fixation for 2–3 days does not adversely affect results.

2. Complete the fixation and embedding process by following steps 2–17 in Method 4.

3. Cut 7–10 μm sections as ribbons on a rotary microtome using a disposable blade.

4. Ribbons are placed on a pool of 20% ethanol in DEPC-treated water on a microscope slide and cut into smaller strips with a clean razor blade if required.

5. The sections are floated out onto distilled water in a clean section bath at 45 °C until any creases disappear.

6. Sections are collected onto TESPA-treated slides (see Method 3), and dried at 48 °C on a hot plate for 2 h, or overnight at 37 °C.

   It is useful to collect at least two adjacent sections per slide; this aids the subsequent identification of positive tissues as they should give identical results on both sections.

7. Store the desiccated slides at 4 °C such that they are free from moisture and sections still give acceptable results months after being cut. Before use allow slides to warm to room temperature.

8. Dewax slides twice in xylene for 10 min each in a fume cupboard.

9. Place in 100% ethanol for 2 min to remove most of the xylene.

10. Re-hydrate sections through the following graded alcohol series: 100% ethanol twice, 95% ethanol, 80% ethanol, 70% ethanol, 50% ethanol and 30% ethanol.

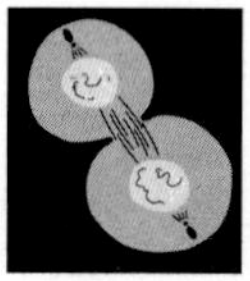

11. Place in 1× saline for 5 min and then in 1× PBS for a further 5 min.

12. Inactivate endogenous peroxidases with 3% (v/v) hydrogen peroxide in PBS for 10 min at room temperature.

13. Wash three times for 5 min in 1× PBS, then once in 10 mM Tris/HCl (pH 7.4).

14. Incubate with proteinase K (20 μg/ml) in 10 mM Tris/HCl (pH 7.4) for 30 min at room temperature.

15. Wash three times, for 5 min each, in PBS/1% (v/v) Triton X-100.

16. Incubate in 1× TdT buffer, 2.5 mM $CoCl_2$, 6 mM dUTP, 4 mM biotin–16-dUTP, 1% Triton X-100 and 0.5 units/ml TdT for 1 h at 37 °C. Ensure that the section is completely covered (100–200 μl spread gently with a piece of parafilm). Incubation is conveniently performed inside a pre-warmed humidified box placed in an oven.

17. Wash in PBS/1% (v/v) Triton X-100 three times, for 10 min each.

18. Incubate for 1 h in a 1:100 dilution in PBS of streptavidin–fluorescein conjugate or streptavidin–horseradish peroxidase conjugate.

19. Wash in PBS/1% (v/v) Triton X-100 three times, for 10 min each.

20. If using the flourescein conjugate, equilibrate and mount the embryos in DABCO mountant and view using epifluorescence. If using the horseradish peroxidase conjugate, wash embryos twice in PBS each for 10 min and then proceed as below.

21. Replace the above with 5 mg/ml DAB in PBS containing 2 μl/ml 30% (v/v) hydrogen peroxide and develop for 5–15 min in the dark.

22. When background staining begins to appear, stop the reaction with several changes of PBS.

23. Equilibrate and mount under a coverslip in 1× PBS/90% (v/v) glycerol.

3.2 Detection of proteins associated with cell death

Great progress has been made in the past decade in determining the molecular events that underlie apoptosis, and many of the proteins that regulate the process have been identified. However, in most cases analysis involving microscopic techniques is either unsuitable or does not provide absolute evidence for an apoptotic event.

Caspases are a family of a dozen or so proteases whose activity is associated with cell death, and their activation is detected by monitoring cleavage of their target proteins [e.g. poly(ADP)-ribose polymerase] *in vitro*. However, since inactive caspases are present in all cells, detection of the enzymes themselves is not a diagnostic tool for apoptosis. Bcl2 is a mitochondria-associated protein that is involved in protection from apoptosis. It functions, in part, by binding and antagonizing the apoptosis-promoting effects of the related protein Bax. However, simple detection of either Bcl2 or Bax in a cell is not informative as to whether or not it is about to become apoptotic. Likewise, detection of the apoptosis-associated transcription factors p53 or Myc is not always associated with apoptotic events in that cell.

The best hope for an immunohistochemical assay for detection of apoptosis lies in the availability of antibodics that distinguish activated and inactive caspases or that can detect the cleavage of their targets. In the latter regard, the availability of a monoclonal antibody that detects only the caspase-cleaved form of cytokeratin 18 (clone M30; Roche) offers the potential to identify cells undergoing apoptosis in tissues that express the cytokeratin — epithelia primarily. Antibodies that detect cleaved (activated) caspases and that are suitable for immunohistochemistry are beginning to appear, e.g. antibodies to active caspases 7 and 9 from Oncogene Research Products. However, these antibodies have not been tested in this laboratory and their species cross-reactivity is currently not available from the supplier.

Work in the author's laboratory is supported by programme grants from the Medical Research Council and Human Frontier Science Program and project grants from the Biotechnology and Biological Sciences Research Council and the Wellcome Trust. I am grateful to my colleagues Robyn Quinlan and Anthony Graham for providing plates B and C for Fig 1.

References

1. Wylie, A.H., Kerr, J.F.R. and Currie, A.R. (1980) Int. Rev. Cytol. **68**, 251–305

2. Graham, A. (1999) Methods Mol. Biol. **97**, 667–672

3. Gaverilli, Y., Sherman, Y. and Ben-Sasson, S.A. (1992) J. Cell Biol. **119**, 493–501.

4. Gratzner, H.G. (1982) Science **218**, 474–475

In situ hybridization to RNA in tissue sections

Mary Gait and Ivor Mason

Chapter 13

1. *In situ* hydridization

In situ hybridization facilitates rapid determination of spatial and temporal gene expression patterns — alternative methods, e.g. immunohistochemistry, are less certain and require the production of suitable antisera. Thus, *in situ* hybridization has become a standard first step in the characterization of the potential function(s) of a newly identified gene. Although it should always be borne in mind that *in situ* hybridization provides no information concerning either translation of the mRNA into protein or the latter's localization, modification or stability. However, *in situ* hybridization remains an extremely powerful tool in enabling the researcher to rapidly predict possible functions for gene products.

Radioactive *in situ* hybridization to tissue sections was pioneered in studies of sea urchin embryos [1] and was then extended to rodent tissue [2,3]. It has recently been partially replaced by non-radioactive approaches on either whole embryos at early developmental stages or, to a lesser extent, on sections of older embryos and adults. The non-radioactive approaches are faster and, in the case of *in situ* hybridizations to whole embryos, ease the spatial interpretation of results. However, the radioactive approach is more sensitive and is the only method by which many low-abundance transcripts are detectable.

Our radioactive *in situ* hybridization protocol [4] (Figs. 1A–1F) is a modification of an original method of Wilkinson and Green [5] and our non-radioactive procedure is a further adaptation of the same methodology [6] (Fig. 1G). Many of the processes are common to all approaches.

- Those undertaking radioactive *in situ* hybridization should follow Methods 1–5 and then Methods 7–10.

- Those undertaking non-radioactive *in situ* hybridization should follow Methods 1–4 and 6, and then Method 11.
- Those combining *in situ* hybridization with immunohistochemistry should follow Methods 1–4 and 6, and then refer to Method 12 for modifications of Method 11.

Method 1. Fixation and embedding of tissue for sectioning

Materials

- Diethylpyrocarbonate (DEPC) (Sigma-Aldrich, Poole, U.K.)
- DEPC-treated water
  - Add DEPC to double distilled water to a final concentration of 0.1% (v/v), shake and then autoclave the solution.

  All solutions used for steps between sectioning tissue and the first post-hybridization wash are prepared in DEPC-treated water; all subsequent steps use solutions prepared in double-distilled water.
- 10× Phosphate-buffered saline (PBS)
  - 1.3 M NaCl, 70 mM Na_2HPO_4, 30 mM NaH_2PO_4. Check that the pH is 7.0 (if required adjust slightly with HCl or NaOH as appropriate). Add DEPC to 0.1% (v/v) and autoclave.
- 4% (w/v) Paraformaldehyde (Fluka, Poole, U.K.) in PBS
  - This should be prepared freshly on the day of use for pre-treatment of sections, but it can be stored for fixation of embryos. Add 16 g of paraformaldehyde (Fluka no. 76240) to 400 ml of 1× PBS in distilled water and heat at 65 °C in a water bath with occasional swirling until it is dissolved (approx. 90 min). Cool to 4 °C prior to use or store in 20 ml aliquots at −20 °C.

> ***Caution*** *Paraformaldehyde should be weighed out in a fume hood while wearing mask and gloves.*

- 10× Saline
  - 8.3% (w/v) NaCl. Add DEPC to a concentration of 0.1% and autoclave.

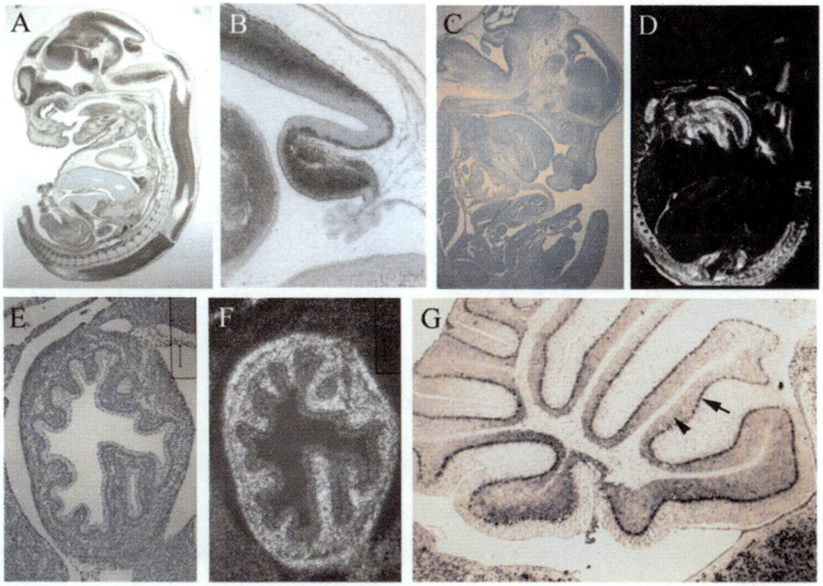

Fig. 1. Radioactive and non-radioactive *in situ* hybridization to tissue sections

(A–F) Radioactive hybridization. (A,B) Some transcripts are expressed so abundantly that the black silver grains are visible using brightfield illumination. The specimen in (A) illustrates this and shows the distribution of transcripts for the actin-binding protein thymosin β_{10} in an embryonic day 11 mouse embryo. (B) Higher magnification of the developing cerebellar region of (A). (C–F) By contrast, deposited silver grains indicating the presence of transcripts for the signalling molecule FGF7 (fibroblast growth factor) cannot be visualized under brightfield illumination (C and E), but are readily seen when the same field is viewed under darkfield illumination (white regions in D and F). Images (D)–(F) are of an embryonic day 16.5 mouse. (C) and (D) are low-power views of the majority of the specimen while (E) and (F) are higher magnification views of the developing duodenum. (G) Non-radioactive in situ *hybridization detects FGF transcripts in the post-natal mouse cerebellum. The arrow shows strong expression in individual Purkinje cells, while the arrowhead shows weaker expression in underlying granule cells.*

- Howard's Ringer
  - 0.12 M NaCl, 0.0015 M $CaCl_2$, 0.005 M KCl. Adjust the pH to 7.2 with very dilute HCl.
- Absolute (100%) ethanol
- Graded ethanols [30%, 50%, 70%, 80% and 95% (v/v) prepared with DEPC-treated water]
- Xylene (AnalaR, Merck BDH, Poole, U.K.)
- Histoclear (National Diagnostics, Hull, U.K.)
- Wax (e.g Paramat pastillated wax, Merck BDH)
- Plastic moulds for wax (R.A. Lamb, Eastbourne, U.K.)

Procedure

1. Dissect embryos or tissues in 1× PBS (mammals) or an appropriate Ringer solution (other vertebrates). Fix in a large volume (at least five times the volume of tissue) of 4% (w/v) parafomaldehyde in PBS at 4°C overnight in a vessel which adequately contains the fumes.

   Smaller embryos or tissue pieces can be fixed for shorter periods of time (4 h minimum) and it is convenient to dechorionate zebrafish after fixation. Embryos and flat tissues can be fixed pinned out in Sylgard-coated Petri dishes (see Chapter 1) to ensure that they do not curl up. Fixation times will vary with the size of the embryo or tissue from a few hours (small embryos) to 12 h (e.g. mice from birth) or even longer. In our experience fixation for 2–3 days does not adversely affect results. Perfusion is the method of choice for adult specimens (see Chapter 6)

2. Incubate in 1× PBS on ice for 30 min (small embryos and tissue pieces) or 1 h (larger material).

   Use glass containers as later treatments will involve high temperatures and/or organic solvents.

Note *Between this stage and until after hybridization of riboprobe there must be no chance of contamination with RNase. RNases are present on human skin so wear gloves, use disposable plastics instead of glass and ideally use pipettemen designated for RNA use only. DEPC-treat all solutions and glassware to inactivate RNases.*

3. Further dissect embryos or tissues if required. Stage the embryos.
4. Remove PBS and replace with 1× saline, incubate on ice for 30 min (small tissue pieces or embryos) or 1 h (larger material).
5. Replace saline with saline/ethanol (1:1, v/v) and incubate for 30 min to 1 h on ice.
6. Replace saline/ethanol with 70% (v/v) ethanol (prepared with DEPC-treated water) and incubate for 30 min to 1 h on ice.
7. Repeat step 6.
8. Replace 70% (v/v) ethanol with 80% (v/v) ethanol and incubate on ice for 30 min.
9. Replace with 95% (v/v) ethanol, and incubate on ice for 30 min to 1 h (larger material) on ice.
10. Replace with 100% ethanol, and incubate on ice for 30 min to 1 h.
11. Place embryos in 100% ethanol for 30 min to 1 h.
12. Replace with 100% ethanol for 30 min (small tissue pieces) to 2 h (larger material).
13. Replace 100% ethanol with xylene or Histoclear in glass bottles and incubate on ice for between 30 min and 1 h.

    We prefer to use Histoclear as it is non-toxic, although some workers find the smell nauseating.
14. Repeat Step 13 with fresh xylene or Histoclear.

15. Pour off xylene or Histoclear and half fill bottles with fresh molten wax (freshly melted wax has more elasticity than wax which has remained molten for days and improves sectioning) and place at 52 °C for between 45 and 90 min.

16. Change the wax twice, incubating each time for 60–90 min.

17. Embed the tissue. Fill moulds with wax (do this on a warm-plate to keep the wax molten), carefully transfer the embryo/tissue to the mould, orient using a hot mounted needle or warmed forceps, and move onto the bench to set. Label one end of the block for reference after the wax is set as the specimen is no longer visible (we insert a piece of card on which details of the specimen and its orientation are written). Allow the wax to set completely; this will take 1–2 h.

Blocks may be stored at 4 °C for several years. Any degradation of RNA during storage is usually due to incomplete penetration of reagents during Steps 4–16. Thus, it is crucial during fixation, processing and embedding in wax prior to sectioning that the reagents completely penetrate and equilibrate within the tissue. As mentioned earlier, times vary according to the size of the embryo/tissue being prepared. The following illustrates modifications which we have used successfully on larger tissues.

- Fix in 4% paraformaldehyde in 1× PBS overnight at 4 °C.
- Wash twice in PBS for 30 min each, and twice in saline for 30 min each. Incubate in 30% ethanol for 1 h, 50% ethanol for 1 h, 70% ethanol for 1 h, 100% ethanol for 3 h and then in fresh 100% ethanol overnight.
- Incubate in fresh 100% ethanol for a further 5 h. Then incubate in three changes of Histoclear (or xylene); the first for 1 h, the second for 3 h and the last overnight.
- Replace the Histoclear with freshly melted (52 °C) wax and incubate at 52 °C in an oven for 3 h, replace the wax with fresh wax and incubate at 52 °C for 5 h and then replace once more and incubate at 52 °C overnight.
- Incubate with two more changes of wax for 3 h each and then embed the tissue.

Method 2. Preparation of slides

Materials

- Washed glass microscope slides and coverslips (Merck BDH)
- 25-Place metal slide racks (R.A. Lamb)
- 400 ml Glass troughs with lids suitable to hold slide racks (R.A. Lamb)
- Concentrated HCl (Merck BDH)
- Acetone (AnalaR, Merck BDH)
- 3-Aminopropyltriethoxysilane (TESPA) (Sigma-Aldrich)
  - □ Use in the fume hood.

Microscope slides need to be washed and silanized prior to collection of sections. Alternatively, pre-treated slides (e.g. Superfrost, Merck BDH) are available commercially.

Procedure

1. Place metal racks of 25 slides in 400 ml of 70% (v/v) ethanol containing 10% (v/v) concentrated HCl, then in 400 ml of distilled water, and finally in 400 ml of 95% (v/v) ethanol, for 10 s each.
2. Wrap the rack of slides in foil to keep free from dust and place the slides at 150 °C for 30 min.
3. Cool the foil-wrapped slides on a bench.
4. Remove the foil and place the rack of slides in acetone containing 2% (v/v) TESPA, then in two changes of acetone and finally in distilled water (not DEPC treated) (for 10 min each). Do this step in the fume cupboard.
5. Dry the slides at 37 °C overnight, again wrapped in foil.
6. Slides are stored individually in foil-wrapped racks (to avoid scratches) at room temperature for up to 3 months.

Method 3. Cutting and mounting sections

Materials

- A suitable rocking microtome

- Disposable blades and holder

> ***Note** Using disposable blades reduces chance of contamination with RNase from a generally-used blade.*

- Bath for sections (well-cleaned with detergent, rinsed well with distilled water and filled with DEPC-treated water)
- Hot plate
- Plastic slide boxes suitable for storing slides, with sections that keep slides well separated.
- Silica gel (Sigma-Aldrich)

Procedure

1. Cut 7–10 μm sections as ribbons on a rotary microtome.
2. Ribbons are placed in a pool of 20% ethanol in DEPC-treated water on a microscope slide and cut into smaller strips with a clean razor blade if required.

   The 20% ethanol assists in removing creases from the sections.
3. Float sections on DEPC-treated water in a clean section bath at 45 °C until creases disappear.
4. Collect onto TESPA-treated slides, and dry at 48 °C on a hot plate for 2 h or overnight at 37 °C.

   We usually collect at least two adjacent sections on a single slide for subsequent identification of hybridizing tissues; identical results should be obtained from both sections.

   If stored desiccated at 4 °C, sections still give acceptable results months after being cut. However, slides should be allowed to warm to room temperature before using them. If unused slides are to be returned to the storage box, place them in a 37 °C oven for 30 min to remove any moisture, and store with a fresh desiccant.

Method 4. Preparation of DNA template for riboprobe synthesis

Materials

- Appropriate restriction endonucleases and manufacturer's buffers (e.g. Roche)
- Klenow polymerase (Roche)
- Agarose (Gibco–BRL, Paisley, U.K.)
- 10× TBE
  - □ 108 g of Tris base, 55 g of boric acid, 9.4 g of EDTA per litre.
- Ethidium bromide (10 mg/ml)
  - □ Stored at 4°C in the dark.

> ***Caution*** *Ethidium bromide is highly carcinogenic.*

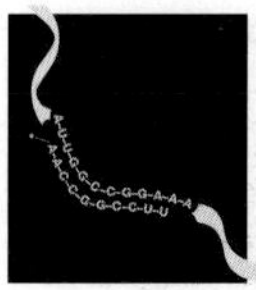

- Phenol (molecular biology grade; Gibco–BRL)
  - □ Melt the phenol at 65 °C in a fume cupboard and equilibrate with 0.1 M Tris/HCl (pH 7.0). Add a few crystals of 8-hydroxyquinoline (Sigma) which will colour the solution yellow. Store in the dark at 4 °C and discard the solution if it begins to turn orange.
- Chloroform (AnalaR; Merck BDH)
- 3 M Sodium acetate (pH 7.0) (adjust pH with acetic acid)
- 100% (v/v) Ethanol
- 70% (v/v) Ethanol

The DNA fragment should have been previously cloned into a plasmid vector so that the prokaryotic RNA polymerase (SP6, T3 or T7) promoters are flanking it. Suitable vectors are available from a variety of manufacturers. In our experience, T7 and T3 RNA polymerases give a better performance on a wide range of templates, compared with SP6 polymerase, which has problems transcribing certain templates.

Restriction enzyme sites should be selected to produce a transcript of 300–1000 bp as larger probe transcripts may increase non-specific background signal. Where possible select enzymes to produce a blunt end or a 5′ overhang, because 3′ overhanging ends can result in 'wrap-around' transcripts due to the polymerase transcribing back along the complementary DNA strand. If there is no alternative to using an enzyme producing a 3′ overhanging end, the following procedure can be used after digestion to 'end polish' generating a blunt end.

Procedure

1. Linearize plasmid DNA (10–20 μg) with a suitable restriction endonuclease using manufacturer's buffer and conditions so that an antisense transcript will be generated.
2. Add 5 units of Klenow polymerase (Roche) per μg of DNA template and incubate at 22 °C for 15 min.
3. Check that digestion is complete by analysing 0.25 μg on a 1% (w/v) agarose gel in 1× TBE containing ethidium bromide (0.1 μg/ml) alongside some uncut plasmid as a negative control. If the digestion is incomplete add more enzyme, incubate at room temperature for a further period and re-analyse by gel electrophoresis.
4. Extract the remainder of the digestion mixture once with phenol, once with phenol/chloroform (1:1, v/v) and twice with chloroform.
5. Precipitate the plasmid DNA by addition of 0.1 volumes of 3 M sodium acetate (pH 7) and 2.2 volumes of ethanol. Incubate on ice for 15 min.
6. Collect the precipitate by centrifugation at 10000 *g* for 10 min, remove the supernatant and wash the pellet with 200 μl of 70% (v/v) ethanol.
7. Allow the last traces of ethanol to evaporate, resuspend the pellet in DEPC-treated water to give a final concentration of 0.5 μg/μl and store at −20 °C.

Method 5. Synthesis of [^{35}S]UTP-labelled riboprobes

Materials

- RNA polymerases and manufacturer's (e.g. Roche) buffers
- 200 mM Dithiothreitol (DTT) (Roche)

- [^{35}S]UTP (>1000 Ci/mmol, Amersham plc, Little Chalfont, U.K.)
- RNasin (Roche)
- RNase-free DNase I (Roche)
- Yeast total RNA (Sigma-Aldrich)
  - Dissolved at a concentration of 10 mg/ml in DEPC-treated water and stored as aliquots at −20 °C.
- 5 M Ammonium acetate (Sigma-Aldrich) dissolved in DEPC-treated water
- Aqueous scintillant, e.g. Aquasol (Dupont NEN, Hounslow, U.K.)

Procedure

1. Bring the buffers, nucleotides, DTT and templates to room temperature.
2. Set up the reaction mixture containing 10 µl of [^{35}S]UTP, 2 µl of 10× transcription buffer, 1 µl of 200 mM DTT, 1 µl of RNasin, 2 µl of a 2.5 mM mixture of nucleoside triphosphates (pH 8.0) (2.5 mM each of CTP, GTP, and ATP) and 1 µl of DEPC-treated water. Mix by gently pipetting up and down.
3. Add 1 µl of the appropriate RNA polymerase and mix gently by pipetting (do not vortex-mix).
4. Add 2 µl (1 µg) of DNA template, mix and spin for 5 s in a microfuge to collect all liquid in the bottom of the tube.
5. Incubate at 37 °C for 1–2 h.
6. Add 10 µl of DEPC-treated water, 2 µl of yeast RNA and 2 µl of RNase-free DNase I, and incubate at 37 °C for 15 min.
7. Add 50 µl of 5 M ammonium acetate (prepared with DEPC-treated water) and 200 µl of ethanol. Incubate on ice for 30 min and spin at 10 000 *g* in a microfuge for 15 min.
8. Remove the supernatant and immediately dissolve the pellet in 30 µl of 10 mM DTT.
9. Repeat step 7.

10. Remove the supernatant and immediately re-dissolve the probe in 20 μl of 100 mM DTT over a period of 15–20 min with occasional vortex-mixing. Obtain a scintillation count for 1 μl in an aqueous scintillant. Calculate probe specific radioactivity in d.p.m./μl per kb of probe length: acceptable probes must be $>10^6$ d.p.m./μl per kb, i.e. specific radioactivity of 1 μl × length of labelled RNA transcript in kb. Store the probes at −70 °C for up to 2 months.

Method 6. Synthesis of non-radioactive riboprobes

Materials

- 10× Digoxygenin (DIG) or fluorescein-12 UTP [fluorescein isothiocyanate (FITC)] ribonucleotide mix (Roche)
- 5× Transcription buffer (usually provided by manufacturer with RNA polymerase — Roche or Promega, Southampton, U.K.)
- 100 mM DTT (Sigma-Aldrich)
- RNasin (Roche)
- RNA polymerase (SP6, T3 or T7 polymerase as appropriate — Roche or Promega)
- 1× TBE (see Method 4)
- Agarose
- Ethidium bromide (10 mg/ml)
- RNase-free DNase I (Roche)
- TE buffer [10 mM Tris/HCl (pH 7.0), 1 mM EDTA; all reagents prepared with DEPC-treated water]
- 4 M LiCl prepared with DEPC-treated water

Procedure

1. Prepare the reaction mixture on ice in the following order: 7 μl of DEPC-treated water, 2 μl of 10× DIG or fluorescein-12 UTP ribonucleotide mix, 4 μl of 5× transcription buffer, 4 μl of 100 mM DTT, 1 μl of RNasin, 1 μl of an appropriate RNA polymerase and 1 μl (1 μg) of linear DNA template.

2. Mix by gently pipetting up and down, and incubate at 37 °C for 2 h.

3. Check that probe has been synthesized by analysing 1 μl of product on a 1% (w/v) agarose gel containing ethidium bromide (0.1 μg/ml). A DNA band and an RNA band of the same intensity should be seen under UV indicating that there is approx. 10 times more DNA then RNA present (ethidium bromide intercalates poorly into RNA).

4. Add 2 μl of RNase-free DNase I and incubate at 37 °C for 15 min (optional).

5. Add 100 μl of TE, 10 μl of 4 M LiCl and 300 μl of ethanol. Precipitate on ice for 10 min (longer incubation and lower temperatures may precipitate unincorporated nucleotides).

6. Spin in a microfuge at 10000 ***g*** for 15 min.

7. Wash the pellet in 70% ethanol, air dry, and resuspend the pellet in 100 μl of TE (made with DEPC-treated water) to give a concentration of approx. 0.1 μg/μl (as estimated from the gel).

8. Add 2 μl of RNasin and store at −20°C.

   A second round of precipitation can be included here, which we find gives significantly reduced background for FITC-labelled probes.

Method 7. Hybridization to tissue sections

Materials

- 4% Paraformaldehyde in PBS (see Method 1)
  - Make up 400 ml fresh on the day of use.
- 100% ethanol, 95% ethanol, 80% ethanol, 70% ethanol, 50% ethanol and 30% ethanol
- Proteinase K (Roche)
  - 20 mg/ml in water stored in single-use, 20 μl aliquots at −20 °C.
- 10× PBS
- 10× Saline
- TE

- 10 mM Tris/HCl (pH 8.0), 1 mM EDTA (pH 8.0) diluted in DEPC-treated water and prepared from concentrated stocks dissolved in DEPC-treated water.

- Triethanolamine (Merck BDH)
- Acetic anhydride (Merck BDH)
- Magnetic stirrer
- Stirring bar
- Radioactively labelled probe
- 20× SSC (3 M NaCl, 0.3 M sodium citrate)
- 1× Denhardt's solution
  - 0.02% Ficoll 400/0.02% polyvinylpyrrolidone/0.002% BSA (all Sigma-Aldrich)
- 50% Formamide
- Hybridization buffer
  - 50% (v/v) Deionized formamide, 0.3 M NaCl, 20 mM Tris/HCl (pH 8.0), 5 mM EDTA (pH 8.0), 10% (w/v) dextran sulphate, 1× Denhardt's solution and 0.5 μg/ml yeast tRNA. Store in aliquots at −70 °C.

Ensure that you have 3.5 litres of DEPC-treated water for dilution of saline etc.

Use designated RNase-free glassware and racks for all of the following procedures, wear gloves throughout, and use DEPC-treated water and solutions.

Procedure

1. Store 400 ml of fresh 4% paraformaldehyde in PBS on ice.

2. De-wax slides twice in xylene for 10 min in a fume cupboard.

   For this and all subsequent steps we use 350 ml of each solution for each incubation; however, this must be sufficient to cover the slides completely. Slides are processed in metal racks that hold up to 25 individual slides in appropriately sized staining troughs. Do not agitate the slides during incubations as it may cause sections to detach.

3. Place in 100% ethanol for 2 min to remove most of the xylene.

4. While performing Step 3, prepare 1.6 litres of $1\times$ PBS and 800 ml of $1\times$ saline.

5. Re-hydrate sections by 'dipping' through the following graded ethanol series: 100% ethanol twice, 95% ethanol, 80% ethanol, 70% ethanol, 50% ethanol and 30% ethanol.

   Dipping simply involves immersion and then withdrawal. Ethanols can be reused up to ten times

6. Place in $1\times$ saline for 5 min and then in $1\times$ PBS for 5 min. Discard used saline and PBS

7. Place in the freshly-made, cold 4% paraformaldehyde in PBS for 20 min in a fume hood.

   Do not discard the fixative as it can be used in Step 12.

8. Meanwhile, prepare 15 ml TE (in DEPC-treated water) (pH 8.0) mixed with 20 μl of proteinase K (20mg/ml) per 25 slides. Also, wash sufficient coverslips of appropriate size in ethanol and leave to air dry in a rack covered in foil.

9. Place slides in $1\times$ PBS twice for 5 min, and then lay flat on a sheet of foil on the bench with the sections uppermost.

10. Cover sections with 300 μl or more of proteinase K (sufficient to cover the sections), and incubate for exactly 7 min at room temperature.

11. Tip off the proteinase K and place slides in a rack in fresh $1\times$ PBS for 5 min.

12. Return the slides to the cold paraformaldehyde (from Step 7) for 5 min in the fume hood.

13. Meanwhile, fill two troughs with DEPC-treated water and place in the fume hood; to the second (acetylation bath), add 6 ml of triethanolamine and 1 ml of acetic anhydride and mix with a small stirring bar on a magnetic stirrer.

14. Dip slides briefly into the DEPC-treated water and then place in the acetylation bath for 10 min with stirring.

    A small enough stir bar or 'flea' will fit beneath standard slide racks and allow stirring in the presence of the latter. This step acetylates slides and sections, and helps to prevent non-specific binding of probe.

15. Place slides in fresh 1× PBS for 5 min and then in fresh 1× saline for 5 min.

16. Dehydrate by 'dipping' through the same ethanol series used in Step 5 starting with 30% ethanol and increasing to two washes with 100% ethanols.

17. Place slides section-side up to air-dry on foil. Add the probe to them on the same day.

18. Dilute the probe in hybridization buffer to give a specific radioactivity of 10^5 d.p.m./μl per kb.

19. Mix the probe by vortex-mixing and spin briefly. Heat to 80 °C for 3 min to remove secondary structures caused by internal hybridization of the probe, allow to cool briefly on ice and then mix again.

20. Place 15–30 μl of probe solution on slide adjacent to sections, spread over sections with a piece of parafilm and cover with a clean coverslip (Step 9) taking care not to introduce any bubbles.

21. Slides are collected on their sides in a rack, which is then placed inside a suitable box with a tightly sealing lid. Include a tissue soaked in 5× SSC/50% formamide to prevent dehydration. Seal the box with suitable tape and hybridize at 55 °C overnight in an oven.

Method 8. Post-hybridization washes

Materials

- 2× SSC
- TNE buffer (0.5 M NaCl, 10 mM Tris/HCl, 5 mM EDTA; pH 8)

- High stringency buffer
  - 50% (v/v) formamide, 2× SSC, 10 mM DTT prepared fresh on day of use using solid DTT or a 100 mM stock of DTT stored at −20 °C.
- RNase A (10 mg/ml) in water (stored at −20 °C)
- Graded ethanols [30%, 50%, 70% and 95% (v/v)] containing 0.3 M (final) ammonium acetate
- Coplin jars (R.A. Lamb)
- Waterbath

  The presence of the ammonium acetate ensures that probe–target-RNA duplexes remain intact during the dehydration process. Duplexes are stable in 100% (v/v) ethanol.

For all procedures after hybridization, use separate glassware and graded ethanol solutions to those used prior to hybridization because RNase I is used during the washing procedures and will potentially contaminate all glassware and re-used solutions.

Procedure

1. Place slides in 4× SSC at room temperature until coverslips fall off (approx. 20 min).

2. Remove coverslips and place slides in fresh 4× SSC for 1 h.

3. Meanwhile, make up 1.6 litres of TNE buffer and 80 ml of high stringency buffer per 25 slides, pre-warming the latter to 50 °C or 65 °C as appropriate (see Step 4 below).

4. Place slides back-to-back in Coplin jars containing high stringency buffer in a 65 °C waterbath (high stringency) or a 50 °C waterbath (low stringency) for 30 min.

   Coplin jars containing 30–40 ml of solution are used at this stage and in Step 8 to reduce the volume of high stringency buffer that is used, as the large amounts of DTT required would otherwise prove very expensive.

5. Transfer slides to a rack and place in 350 ml of TNE buffer for 10 min at 37 °C and then repeat this wash.

6. Place slides in a third trough containing 400 ml of pre-warmed TNE buffer and 800 μl of 10 mg/ml RNase A for 30 min at 37 °C.
7. Wash slides for 10 min in fresh TNE buffer at 37 °C.
8. Wash slides for 30 min in Coplin jars containing pre-warmed high stringency buffer at 65 °C (or at an appropriate temperature).
9. Wash for 15 min in 2× SSC and then for 15 min in 0.1× SSC (both washes at room temperature).
10. Dehydrate through graded ethanols (from 30% to 100%) containing 2.5 M ammonium acetate (to prevent ‘melting’ of the probe from target RNAs) by dipping, as described in Method 7.
11. Air dry in a rack loosely wrapped in foil for 60 min minimum. The slides may be left overnight to dry.

Method 9. Autoradiography

Materials

- Emulsion (K-S no. 1355127, Ilford Photographic, Knutsford, U.K.)
- Glycerol (Merck BDH)
- Two-place plastic slide mailers (R.A. Lamb)
- Aluminium foil
- 50 ml Falcon tube
- Glass sheet
- Water bath
- Silica gel packets
- Water
- Glycerol

All steps are performed in a darkroom using a red safe light.

Procedure

1. Cool a foil-wrapped glass sheet on a tray of ice in a darkroom.

2. Place approx. 15 ml of emulsion ribbons into a 50 ml Falcon tube and melt them in a 42 °C water bath.
3. Decant 8 ml of the melted emulsion into a slide mailer and place at 42 °C.
4. Prepare 8.8 ml of water containing 180 μl of glycerol and add to the melted emulsion. Mix gently with a glass pipette to expel bubbles and warm to 42 °C in the water bath for 10 min.
5. Dip slides briefly, and at a uniform speed, into the emulsion, wipe the back (non-section side) of each and lay the section side upon the cold tray for 10 min in order to set. If bubbles or streaks are visible then re-dip the section.
6. Place the slides in a box in complete darkness for 2 h, to harden the emulsion, then transfer to a sealed light-tight box containing silica gel packets, and store at 4 °C for a few days to 2 weeks, depending upon abundance of target transcript.

Method 10. Developing and mounting after autoradiography

Materials

- D19 developer (Kodak)
  - Dissolve 64 g of D19 developer in 400 ml of water. This can be used several times on the day of preparation, but it should be discarded after use.
- Stop solution [350 ml of 1% (v/v) acetic acid/1% (v/v) glycerol]
- Fixing solution [350 ml of 30% (w/v) sodium thiosulphate]
- 0.25% Toluidine Blue (Sigma-Aldrich) filtered through a Whatman general purpose filter.
- Distilled water
- Graded ethanols [30%, 50%, 80%, 95%, 100% (v/v)]
- Depex mounting medium (Merck BDH).

Unless otherwise stated, all procedures are carried out in the darkroom using a red safe-light.

Procedure

1. Add 350 ml portions of fresh developer, stop solution and fixing solution to individual troughs. Allow the temperature of all solutions to reach 20°C and place them in the darkroom.
2. Remove slides from the fridge and allow them to equilibrate to room temperature.
3. Place a rack of slides in the developer for exactly 150 s, then place them directly into the stop solution for 1 min and into the fixing solution for 5 min.
4. Wash in troughs of distilled water twice, each for 10 min. Slides can now be removed from the darkroom.
5. For counterstaining, place each slide in 50 ml of freshly filtered 0.25% (w/v) Toluidine Blue for exactly 1 min. Dehydrate by rapid dipping through a graded series of ethanols (30–100%), washing twice in 100% ethanol.
6. Clear with xylene (Histoclear cannot be used here as it leaves an oily film on the slide) and mount with coverslips under Depex.

Method 11. Hybridization and detection of non-radioactive riboprobes

Materials

- TBST buffer
  - 50 mM NaCl, 100 mM Tris/HCl (pH 9.5), 0.1% (v/v) Tween-20.
- Heat-inactivated sheep serum (Sigma-Aldrich) (stored at −20 °C; centrifuge before use)
- Anti-DIG alkaline phosphatase-conjugated antibody (F_{ab} fragments) (Roche)
- Anti-FITC alkaline phosphatase-conjugated antibody (F_{ab} fragments) (Roche)
- Nitro-Blue Tetrazolium solution (NBT) (Roche)
- 5-Bromo-4-chloro-indolyl phosphate (BCIP) solution (Roche)
- Proteinase K (20 mg/ml)

- NTMT
  - 100 mM NaCl, 100 mM Tris/HCl (pH 9.5), 50 mM $MgCl_2$, 0.1% Tween-20 (make fresh on the day of use from a concentrated stock of each individual component).
- Fast Red TR/naphthol AS-MX (Sigma-Aldrich)
- Glycergel
- Triethanolamine
- Acetic anhydride
- DEPC-treated water (a minimum of 3.5 litres)
- Graded ethanols (100%, 95%, 80%, 70%, 50% and 30%)
- TE buffer (pH 8.0)
  - 10 mM Tris/HCl (pH 8.0), 1 mM EDTA; prepared with DEPC-treated water.
- Hybridization buffer (see Method 7)
- 20× SSC
- Formamide
- PBS
- Saline solution
- RNase A
- Aluminium foil
- Magnetic stirrer and stirring bar
- Hybridization oven
- Coplin jar

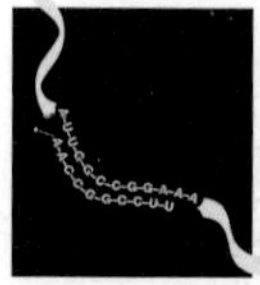

> ***Note*** *Use designated RNase-free glassware and racks for all of the following procedures, wear gloves throughout, and use DEPC-treated water and solutions.*

Procedure

1. Make up 400 ml of fresh 4% paraformaldehyde in PBS and store on ice (see Method 1).
2. De-wax slides in xylene twice for 10 min in a fume cupboard.

   A volume of 350 ml of each solution is recommended for each incubation; however, this must be sufficient to cover the slides completely. Slides are processed in metal racks holding up to 25 individual slides in appropriately sized staining troughs. Do not agitate the slides during incubations as it may cause sections to detach.

3. Place in 100% ethanol for 2 min to remove most of the xylene.
4. While performing Step 3 prepare 1.6 litres of 1× PBS and 800 ml of 1× saline.
5. Re-hydrate sections by 'dipping' through the following graded ethanol series: 100% ethanol (twice), 95% ethanol, 80% ethanol, 70% ethanol, 50% ethanol and 30% ethanol.

   Dipping simply involves immersion and then withdrawal. Ethanols can be reused up to ten times.
6. Place the slides in 1× saline for 5 min and then in 1× PBS for a further 5 min. Discard used saline and PBS.
7. Add the slides to the freshly-made, cold 4% paraformaldehyde in PBS for 20 min in the fume hood. This solution can be reused in Step 13.
8. Prepare 15 ml of DEPC-treated TE (pH 8.0) and add 20 μl of proteinase K per 25 slides.
9. Wash a sufficient number of coverslips of appropriate size in ethanol and leave to air dry in a rack covered in aluminium foil.
10. Place the slides in 1× PBS twice (each for 5 min), and then lay flat on a sheet of foil on the bench with the sections facing up.
11. Cover sections with 300 μl or more of proteinase K solution (from Step 8) (sufficient to cover the sections) and incubate for exactly 7 min at room temperature.
12. Tip off the proteinase K and place the slides in a rack in fresh 1× PBS for 5 min.
13. Return the slides to the cold paraformaldehyde for 5 min in the fume hood.

14. Fill two troughs with DEPC-treated water and place in the fume hood. To the second trough (acetylation bath), add 6 ml of triethanolamine and 1 ml of acetic anhydride and mix with a small stir-bar on a stirrer.

15. Dip slides briefly into the DEPC-treated water and then place them in the acetylation bath for 10 min. The acetylation mixture should be a stirred.

    A small stir-bar or 'flea' will fit beneath standard slide racks to allow stirring.

    This step acetylates slides and sections and helps prevent non-specific binding of probe.

16. Place slides in fresh 1× PBS for 5 min and then in fresh 1× saline for 5 min.

17. Dehydrate by 'dipping' through the same ethanol series used in Step 5 starting with 30% ethanol and increasing to the two dips in 100% ethanol.

18. Place slides section-side up to air-dry on foil. Probe should be added on the same day.

19. Dilute the probe in hybridization buffer to give a final concentration of 1 μg/ml.

20. Vortex-mix the probe, spin briefly, heat to 80 °C for 3 min (removes secondary structures due to internal hybridization of the probe), allow to cool briefly on ice and then mix again.

21. Place 15–30 μl of probe solution on slide adjacent to sections, spread over sections with a piece of parafilm and cover with a clean coverslip, taking care not to introduce any bubbles.

22. Slides are collected on their sides in a rack, which is then placed inside a suitable box with a tightly-sealing lid. Include a tissue soaked in 5× SSC/50% formamide to prevent dehydration. Seal the box with suitable tape and hybridize at 55 °C overnight in an oven.

23. Place slides in 4× SSC at room temperature until coverslips fall off (approx. 20 min).

24. Remove coverslips and place slides in fresh 4× SSC for 1 h.

25. Make up 1.6 litres of TNE buffer and 80 ml of high stringency buffer (see Method 8) per 25 slides. Pre-warming the latter to 50 °C or 65°C as appropriate.

26. Place the slides back-to-back in Coplin jars containing high stringency buffer in a waterbath at 65 °C (high stringency) or 50 °C (low stringency) for 30 min.

27. Transfer slides to a rack and place it in 350 ml of TNE buffer for 10 min at 37 °C. Repeat this wash.

28. Place the slides in a third trough containing 400 ml of pre-warmed TNE buffer and 800 μl of 10 mg/ml RNase A for 30 min at 37 °C.

29. Wash the slides for 10 min in fresh TNE buffer at 37°C.

30. Wash the slides for 30 min in Coplin jars containing pre-warmed high stringency buffer at 65 °C (or at an appropriate temperature; see Step 26).

31. Wash the slides for 15 min in 2× SSC and then for 15 min in 0.1× SSC, both at room temperature.

32. Wash three times (each for 5 min) with TBST.

33. Pre-block sections with 20% (v/v) sheep serum in TBST for 30 min

34. Incubate sections with 1/2000 dilution of anti-DIG alkaline phosphatase antibody in 10% (v/v) sheep serum/TBST, for 2 h at room temperature.

35. Wash three times for 20 min with TBST.

36. Change TBST hourly and continue to wash with TBST throughout the day, with a final overnight wash.

37. Wash three times for 15 min with NTMT.

38. Incubate with NTMT containing NBT and BCIP (4.5 μl/ml and 3.5 μl/ml, respectively, of stock solutions) in the dark until the desired reaction product intensity is achieved.

39. Stop the reaction by washing in TBST. Refix in 4% paraformaldehyde/1× PBS for 1 h.

40. Mount under coverslips in glycergel.

Method 12. Combining immunohistochemistry with *in situ* hybridization to tissue sections

Procedure

1. Follow Methods 1–3 to prepare sections, Methods 4 and 6 to prepare a riboprobe and then Method 11(Steps 1–39) to detect the mRNA of interest).

   The proteinase K step in Method 11 (Step 11) may cleave the antigenic site that is recognized by the antibody. It is therefore suggested that a second set of slides are treated with a mixture of detergents instead of proteinase K.

   Substitute the proteinase K step with two 20 min washes in a detergent mix [1% IGEPAL, 1% SDS, 0.5% deoxycholate, 50 mM Tris/HCl (pH 8), 1 mM EDTA, 150 mM NaCl].

2. Proceed with immunohistochemistry. Wash slides with TBST and then use an immunohistochemistry protocol that has previously been optimized for wax sections, to detect antigen–antibody complexes with an alkaline phosphatase secondary antibody and Fast Red substrate in NTMT (pH 8).

3. Stop the reaction by washing in TBST. Refix in 4% paraformaldehyde/1× PBS for 1 h.

4. Mount under coverslips in glycergel.

Work in I.M.'s laboratory is supported by programme grants from the Medical Research Council and Human Frontier Science Program and project grants from the Biotechnology and Biological Sciences Research Council and the Wellcome Trust.

References

1. Cox, K.H., Deleon, D.V., Angerer, L.M. and Angerer, R.C. (1984) Dev. Biol. **101**, 485–502

2. Wilkinson, D.G., Bailes, J.A. and McMahon, A.P. (1987) Cell **50**, 79–88

3. Wilkinson, D.G., Bhatt, S. and McMaohon, A.P. (1989) Development **105**, 131–136

4. Mason, I.J., Fuller-Pace, F., Smith, R. and Dickson, C. (1994) Mech. Dev. **45**, 15–30

5. Wilkinson, D. and Green, J. (1990) in Postimplantation Mammalian Embryos: A Practical Approach (Copp, A.J. and Cockroft, D.L., eds), pp. 155–171, IRL Press, Oxford

6. Beer, H.-D., Vindevoghel, L., Gait, M.J., Revest, J.-M., Duan, R., Mason, I., Dickson, C. and Werner, S. (2000) J. Biol. Chem. **275**, 16091–16097

In situ hybridization and sectioning of whole embryos and tissues

Carol Irving and Ivor Mason

Chapter 14

1. Introduction

The ability to visualize regions of gene expression in whole embryos or tissues provides a powerful tool that assists in understanding the spatial and temporal relationships between the proteins that these genes encode (see e.g. Fig. 1A). Moreover, it frequently provides the first clues as to potential functions of a novel gene. Performing *in situ* hybridization before sectioning is advantageous because large sample numbers can be processed rapidly and spatial aspects of gene expression are more readily appreciated. Sites of specific interest can subsequently be sectioned for visualization at single cell resolution.

Recently, wholemount *in situ* hybridization has become a common technique and a number of protocols are therefore available, with modifications for different organisms. Here we describe two protocols that can be used for a variety of species. The first approaches are suitable for hybridization to embryos or tissues with probes derived from DNA sequences either from the same species or from other species (Fig. 1A; Methods 2 and 3). This protocol is a modified version of Wilkinson's method [1]. It also includes a method for 'two-colour' *in situ* hybridization, which allows the simultaneous expression of two transcripts to be examined (Fig. 1B). A later procedure is modified from that of Henrique et al. [2] and provides both stronger pre-blocking and a more stringent hybridization (Method 4). It is therefore useful for probes where background and cross-reactivity appear problematic. As tissue or embryo sizes (volume) increase, non-specific 'background' hybridization becomes a problem. This can be avoided by cutting the specimen into a number of 'thick' slices prior to hybridization; a procedure for this is provided (Method 5). In addition, we also provide a method that couples both single and double *in situ* hybridization with wholemount immunohistochemistry (Fig. 1C; Section 5).

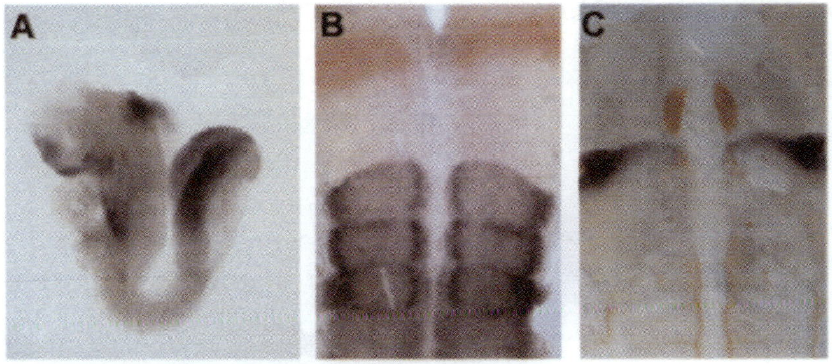

Fig. 1. *In situ* hydridization of whole embryos

(A) In situ hybridization using an Fgf8 antisense RNA probe on an 8.5-day mouse embryo. (B) Double in situ hybridization using DIG-labelled (Hoxa2, blue) and fluorescein-labelled (Fgf8, red) antisense probes on a 3-day chick embryo, from which the neural tube has been flatmounted. (C) Combined in situ hybridization (Fgf8, blue) and immunohistochemistry (Isl1/2 antibody, brown) of a 4-day chick embryo, from which the neural tube has been flatmounted.

2. Preparation of a riboprobe

This procedure is the same regardless of which hybridization technique is subsequently utilized. Probes are generated by *in vitro* transcription of a DNA template, which should be cloned into a vector containing appropriate polymerase initiation sites. Suitable vectors include the pGEM series (Promega) and pBluescript series (Stratagene), which contain T7, T3 and SP6 polymerase promoters. Probes are complementary to the target mRNA (i.e. an antisense strand is synthesized) and are labelled by the incorporation of hapten-conjugated nucleotides [typically digoxygenin (DIG) gives the best results].

We find that the quality of probe preparation is the most important determinant of strong signals with low background. Plasmids should be linearized to completion at the 5′-end of the cDNA, as inclusion of plasmid sequences in the probe increases non-specific binding and hence background. The ideal probe length is between 300 and 1000 bp. Larger transcripts frequently give a non-specific background signal (probably due to trapping of probe).

Method 1. Preparation of a riboprobe

Materials

- Appropriate restriction endonucleases and manufacturer's buffers (e.g. Roche, Welwyn Garden City, U.K.)
- 10× TBE
    - 0.89 M Tris/borate (0.89 M boric acid, 0.02 M EDTA; 108 g of Tris base, 55 g of boric acid, 9.4 g of EDTA per litre).
- Ethidium bromide (10 mg/ml)
    - Stored at 4 °C in the dark.

Caution *Ethidium bromide is a potent carcinogen.*

- Nuclease-free agarose (e.g. Gibco–BRL, Paisley, U.K.)
- Phenol (molecular biology grade; Gibco–BRL)
    - Melt at 65 °C in a fume cupboard and equilibrated with 0.1 M Tris/HCl (pH 7) followed by the addition of a few crystals of 8-hydroxyquinoline (Sigma-Aldrich), which will colour the solution yellow. Store the phenol in the dark at 4 °C; discard if the solution begins to turn orange.
- Chloroform (AnalaR; Merck BDH; Poole, U.K.)
- 3 M Sodium acetate (pH adjusted to 7.0 with acetic acid) (Sigma-Aldrich)
- 100% (v/v) Ethanol (Merck BDH)
- 70% (v/v) Ethanol
- Diethylpyrocarbonate (DEPC) (Sigma-Aldrich)
    - To DEPC-treat water or an aqueous solution, add DEPC to double distilled water to a concentration of 0.1% (v/v), shake well and autoclave.

All solutions used for steps between sectioning tissue and the first post-hybridization wash are prepared in DEPC-water, all subsequent steps use solutions prepared in double-distilled water.

- 10× DIG or fluorescein-12 UTP [fluorescein isothiocyanate (FITC)] ribonucleotide mix (Roche)
- 5× Transcription buffer (usually provided by manufacturer with RNA polymerase, e.g. Roche or Promega, Southampton, U.K.)
- 100 mM dithiothreitol (DTT) (Roche)
- RNase inhibitor
- RNA polymerase (SP6, T3 or T7 polymerase as appropriate; e.g. Roche or Promega)
- RNase-free DNase I (Roche)
- TE buffer [10 mM Tris/HCl (pH 7.0), 1 mM EDTA; all reagents prepared with DEPC-treated water)
- 4 M LiCl prepared with DEPC-treated water

Linearization of the DNA template

The DNA fragment should have been cloned previously into a plasmid vector so that prokaryotic RNA polymerase (SP6, T3 or T7) promoters flank it. Suitable vectors are available from a variety of manufacturers. In our experience, T7 and T3 RNA polymerases give the better performance than SP6 polymerase on a wide range of templates; the latter has problems transcribing certain templates.

1. Linearize plasmid DNA (10–20 μg) with a suitable restriction endonuclease using manufacturer's buffer and conditions such that an antisense transcript will be generated.

2. Check that digestion is complete by analysing 0.25 μg of plasmid on a 1% (w/v) agarose gel in 1× TBE containing ethidium bromide (0.1 μg/ml), alongside some uncut plasmid as a negative control. If the digestion is incomplete add more enzyme, incubate for a further period and re-analyse by gel electrophoresis.

3. Extract the remainder of the digestion mixture once with phenol, once with phenol/chloroform (1:1, v/v) and twice with chloroform. In each

instance add an equal volume of the organic liquids to the digest, vortex-mix for 10 s, centrifuge at 10000 *g* for 5 min and remove and discard the non-aqueous phase.

4. Precipitate the plasmid DNA by addition of 0.1 volumes of 3 M sodium acetate (pH 7.0) and 2.2 volumes of ethanol. Incubate on ice for 15 min.
5. Collect the precipitate by centrifugation at 10000 *g* for 10 min, remove the supernatant and wash the pellet with 200 μl of 70% (v/v) ethanol.
6. Allow the last traces of ethanol to evaporate, resuspend the pellet in DEPC-treated water to give a final concentration of 0.5 μg/μl and store at −20 °C.

Preparation of DIG- or fluorescein-labelled riboprobes

1. Prepare a reaction mixture containing 7 μl of DEPC-treated water, 2 μl of 10× DIG or fluorescein-12 UTP (FITC) ribonucleotide mix, 4 μl of 5× transcription buffer, 4 μl of 100 mM DTT, 1 μl of RNase inhibitor, 1 μl of an appropriate RNA polymerase and 1 μl (1 μg) of linear DNA template. The reaction mixture should be prepared on ice.
2. Mix by gently pipetting up and down, and incubate at 37 °C for 2 h.
3. Check that the probe has been synthesized by analysing 1 μl of product on a 1% (w/v) agarose gel in 1× TBE containing ethidium bromide (0.1 μg/ml). A DNA band and an RNA band of the same intensity should be seen under UV indicating that approx. 10 μg of RNA has been synthesized (ethidium bromide intercalates poorly into RNA).
4. Add 2 μl of RNase-free DNase I and incubate at 37 °C for 15 min (optional).
5. Add 100 μl of TE (made with DEPC-treated water), 10 μl of 4 M LiCl and 300 μl of ethanol. Precipitate on ice for 10 min (longer incubation and/or lower temperatures may precipitate unincorporated nucleotides).
6. Spin in a microfuge at 10000 *g* for 15 min. Wash the pellet in 70% ethanol, air-dry for 10 min, then resuspend pellet in 100 μl of TE (made with DEPC-treated water) to give a concentration of approx. 0.1 μg/μl as estimated from the gel.

7. Add 1 μl of RNasin and store at −20 °C.

   A second round of precipitation can be included here, which we find gives significantly reduced background for FITC-labelled probes.

3. Fixation and prehybridization of specimens

The following points apply to all methods in Section 3.

- Perform all steps at room temperature for 5 min, with rocking or slow rotation unless otherwise stated. For larger tissues increase times to 10–15 min.
- To avoid mechanical damage to embryos, leave a small quantity of liquid over them between each step.
- To avoid degradation of target mRNAs, all reagents must be free from RNase contamination until after hybridization. Wear gloves, use disposable plastics and treat solutions with DEPC where possible.

Method 2. Reduced stringency hybridizations: fixation

Materials

- 10× Phosphate-buffered saline (PBS)
  - 1.3 M NaCl, 70 mM Na_2HPO_4, 30 mM NaH_2PO_4. Check that the pH is 7.0 (if required adjust slightly with dilute HCl or NaOH as appropriate). Add DEPC to a concentration of 0.1% (v/v) and autoclave.
- MEMFA buffer
  - 4% Formaldehyde, 0.1 M 3-(*N*-morpholino)-2-hydroxypropanesulphonic acid (pH 7.4), 2 mM EGTA, 1 mM $MgSO_4$ (all Sigma-Aldrich).
- PBT
  - 1× PBS, 0.1% (v/v) Tween-20
- Methanol (Merck BDH)
- Proteinase K (Roche)
- Glutaraldehyde [25% (w/v), electron microscopy grade, Merck BDH]

- 4% (w/v) Paraformaldehyde (Fluka, Poole, U.K.) in 1× PBS
  - This must be prepared freshly on the day of use for pre-treatment of sections, but can be stored for the fixation of embryos. Add 16 g of paraformaldehyde (Fluka no. 76240) to 400 ml of 1× PBS in distilled water and heat at 65 °C in a water bath with occasional swirling until dissolved (about 90 min). It is cooled to 4 °C prior to use or stored in 20 ml aliquots at −20 °C.

Caution *Formaldehyde must weighed out in a fume hood while wearing mask and gloves.*

Procedure

1. Dissect out embryos in PBS and open cavities.

   Reagents may become trapped inside structures with a large lumen (e.g. heart and neural tube) and give high levels of background. Therefore, cavities should be opened by making a hole with a sharpened tungsten needle.

2. Fix in a large volume of MEMFA at 4 °C for 4 h to overnight. For this and subsequent procedures up to the hybridization step, use an excess (e.g. 20-fold by volume) of solution over embryo tissue. Tissue can be stored at this point for 1 month.

3. Wash twice in PBT.

4. Dehydrate embryos by washing in the following series of methanols: 25% (v/v), 50% (v/v) and 75% (v/v) methanol in PBT. Finally, wash twice with 100% methanol for 20 min each. Embryos can be stored at this point at −20 °C.

   Some authors claim that indefinite storage of material is possible at this point. However, in our experience storage for more than a month gives poor results.

5. Rehydrate embryos by washing with the methanol series in PBT for 10 min (75%, 50% and 25% methanol in PBT). Finally, wash twice with PBT for 10 min each.
6. Treat with 10 mg/ml proteinase K in PBT for 5–15 min. The length of treatment should be optimized depending on the size of the embryo/tissue.

   The optimum time for proteinase K treatment will vary for each new batch of proteinase. Treatment for 10 min at room temperature is a good starting point for small embryos: mouse, 8.5 days *post coitum*; chick, 10 somites; *Xenopus*, up to stage 14. Incubation times can be increased for larger embryos. Zebrafish require shorter incubation times (5–10 min). Over-exposure to proteinase K results in poor signals or disintegration of embryos.
7. Rinse twice with PBT (gently, as embryos are fragile until fixed).
8. Refix embryos using 0.2% (w/v) gluteraldehyde/4% (w/v) paraformaldehyde in PBT for 20 min.
9. Wash three times with PBT.

Method 3. Reduced stringency hybridizations: prehybridization, hybridization, washing and immunohistochemical detection of probe

Embryos are incubated with probe, then washed at moderate stringency to remove any unhybridized probe before immunohistochemical detection using an anti-DIG antibody conjugated to alkaline phosphatase, which is visualized using a chromogenic substrate that forms a purple precipitate at sites of bound probe.

Materials

- 20× SSC
  - 3 M NaCl, 0.3 M sodium citrate; adjust the pH to 4.5 with citric acid.

- Prehybridization solution
  - 50% (v/v) Deionized formamide, 5× SSC, 50 mg/ml yeast tRNA (Sigma-Aldrich), 1% (w/v) SDS, 50 mg/ml heparin (Sigma-Aldrich).
- Hybridization solution (prehybridization solution containing riboprobe)
- Solution 1
  - 50% (v/v) Deionized formamide, 5× SSC, 1% (w/v) SDS.
- Solution 2
  - 0.5 M NaCl, 10 mM Tris/HCl (pH 7.5), 0.1% (v/v) Tween-20.
- Solution 3
  - 50% (w/v) Formamide, 2× SSC.
- RNase A (20 mg/ml stock stored at −20 °C; Roche)
- TBST
  - 50 mM NaCl, 100 mM Tris/HCl (pH 9.5), 0.1% (v/v) Tween-20
- Heat-inactivated sheep serum (Sigma-Aldrich) (stored at −20 °C; centrifuged before use)
- Anti-DIG alkaline phosphatase-conjugated antibody F_{ab} fragments (Roche)
- Anti-FITC alkaline phosphatase-conjugated antibody F_{ab} fragments (Roche)
- Nitro-Blue Tetrazolium solution (NBT) (Roche)
- 5-Bromo-4-chloro-3-indolyl phosphate (BCIP) (Sigma-Aldrich B-6777) (50 mg/ml in dimethylformamide)
- NTMT
  - 100 mM NaCl, 100 mM Tris/HCl (pH 9.5), 50 mM $MgCl_2$, 0.1% Tween-20 (make fresh on day of use from concentrated stock of each individual component).
- Fast Red TR/Naphthol AS-MX (Sigma-Aldrich)
- *p*-Iodonitrotetrazolium Violet (Sigma-Aldrich 18377) (5 mg/ml in water, filtered)

Procedure

1. Remove PBT that was used to wash the fixed embryos and rinse them once with pre-warmed prehybridization solution.

2. Replace with fresh prehybridization solution and incubate at 65–70 °C for a minimum of 30 min or a maximum overnight incubation.

3. Incubate embryos with pre-warmed hybridization solution containing approx. 1 µg/ml of riboprobe (usually 10 µl of probe to 1 ml of prehybridization solution). For double *in situ* assays, add both DIG- and FITC-labelled probes simultaneously. Incubate at 65–70 °C overnight.

4. Wash embryos twice with pre-warmed Solution 1 for 30 min each at 65–70 °C. For all washes use an excess of wash solution (e.g. 20-fold) over the volume of the embryonic tissue.

5. Wash with a mixture of Solution 1 and Solution 2 (1:1, v/v) for 10 min at 65–70 °C

   Steps 5–8 are only necessary for very closely related genes (see note at end of Step 7).

6. Wash three times with Solution 2.

7. Wash twice with 100 mg/ml RNase A in Solution 2 for 30 min at 37 °C.

> ***Note** RNAse A treatment should be omitted if probes are derived from DNA sequences of a different species, and can usually be omitted anyway. However, it is necessary in order to distinguish between transcripts derived from closely related members of a multigene family. For example, expression of transcripts of the vertebrate engrailed genes, En-1 and En-2, can only be distinguished if RNase treatment is included (C. Logan, personal communication). Its inclusion probably also reduces overall background levels.*

8. Wash with Solution 2 and then with Solution 3.

9. Wash twice at 65–70 °C with pre-warmed Solution 3 for 30 min each.

10. Wash three times with TBST.

11. Pre-block embryos with 20% (v/v) sheep serum in TBST for 2 h.

12. Incubate embryos with 1:2000 dilution of anti-DIG alkaline phosphatase-conjugated antibody in 10% (v/v) sheep serum in TBST, overnight at 4 °C, with rocking.

13. Wash three times for 15 min with TBST.

14. Change TBST every hour throughout the day and include a final overnight wash.

15. Wash three times for 15 min with NTMT.

16. Incubate in NTMT containing 0.25 μg/ml NBT and 0.175 μg/ml BCIP in the dark until the desired reaction product intensity is achieved.

    For *in situ* hybridizations using two probes, proceed directly to Step 18.

17. For *in situ* hybridizations using a single probe, stop the reaction by washing thoroughly in TBST. Refix in 4% paraformaldehyde/1× PBS for 1 h prior to storing or sectioning.

    If embryos are to be sectioned, we have found that it is best to 'over-develop' the apparent level of signal on whole embryos in order to get sufficient intensity on sections.

18. For *in situ* hybridizations using two probes, rinse several times with TBST, then inactivate alkaline phosphatase by incubating embryos in TBST at 70 °C for 30–40 min.

19. Wash three times with TBST.

20. Pre-block embryos again with 20% sheep serum in TBST for 60–90 min.

21. Incubate with a 1:2000 dilution of anti-FITC alkaline phosphatase-conjugated antibody in TBST/10% sheep serum overnight at 4 °C with rocking.

22. Repeat Steps 13 and 14.

23. Wash three times for 15 min with NTMT (pH 8.0) [NTMT made with 100 mM Tris/HCl (pH 8)].

24. Incubate with NTMT containing Fast Red TR/Naphthol AS-MX. Incubate in the dark until desired reaction product intensity is achieved (usually two to three times the amount of time required for DIG probes). We find that best results are achieved using Fast Red at half the recommended concentration. Stop the reaction by washing thoroughly in

TBST. Refix in 4% paraformaldehyde/1× PBS for 1 h prior to storing or sectioning.

While Fast Red is the standard 'second substrate' used in two-colour protocols, it often turns negative tissue a yellow colour in prolonged staining reactions. We have recently used an alternative 'second substrate', *p*-Iodonitrotetrazolium Violet (Sigma 18377), which produces an orange product without any apparent coloration of negative tissues [3]. It can be used as follows:

(i) Replace Step 23 with washes in NTMT (pH 9.5).

(ii) Incubate in substrate solution prepared fresh from stocks (substrate solution: 0.188 mg/ml BCIP, 0.2 mg/ml *p*-Iodonitrotetrazolium Violet).

(iii) Incubate embryos in substrate solution until a satisfactory orange colour has appeared. Stop the reaction by washing thoroughly in TBST. Refix in 4% paraformaldehyde/1× PBS for 1 h prior to storing or sectioning.

25. Stop the reaction by washing thoroughly in TBST. Refix in 4% paraformaldehyde/1× PBS for 1 h prior to storing or sectioning.

Method 4. High stringency protocol: prehybridization through to detection of hybridized probe

Materials

- Prehybridization solution
  - 50% (v/v) Formamide, 1.3× SSC, 5 mM EDTA (pH 8.0), 50 mg/ml yeast RNA, 0.2% (v/v) Tween-20, 0.5% CHAPS, 100 mg/ml heparin.
- Hybridization solution (prehybridization solution containing 1 μg/ml of riboprobe)
- Solution A
  - 50% Formamide, 1× SSC, 0.1% Tween-20.
- MABT
  - 100 mM Maleic acid, 150 mM NaCl, 0.1% Tween-20, (final pH 7.5)
- Boehringer blocking reagent (Roche)
- NTMT (see Method 3)

- NBT (see Method 3)
- BCIP (see Method 3)
- 4% Paraformaldehyde (see Method 1)
- PBS (see Method 1)

Procedure

1. Perform fixation and pretreatments exactly as in Method 2.
2. Remove PBT and rinse embryos once with pre-warmed prehybridization solution.
3. Replace with fresh prehybridization solution and incubate at 65–70 °C for a minimum of 30 min to a maximum overnight incubation.
4. Incubate embryos with pre-warmed hybridization solution containing approx. 1 μg/ml of riboprobe.
5. Rinse twice with pre-warmed hybridization mixture.
6. Wash twice with pre-warmed Solution A for 20 min each at 65–70 °C.
7. Wash with pre-warmed 1:1 (v/v) mixture of Solution A/MABT for 10 min at 65–70 °C.
8. Rinse three times with MABT, then wash twice with MABT for 30 min each.
9. Pre-block the embryos in 20% sheep serum/1× Boehringer blocking reagent in MABT for 2 h.
10. Incubate embryos with a 1:2000 dilution of anti-DIG or anti-FITC alkaline phosphatase-conjugated antibody in 10% (v/v) sheep serum/1× Boehringer blocking reagent in MABT, overnight at 4 °C with rocking.
11. Wash three times with MABT, for 15 min each.
12. Change MABT every hour, and continue to wash with MABT throughout the day, with a final overnight wash.
13. Wash three times for 15 min with NTMT (pH 9.5).

14. Incubate with NTMT containing 0.25 μg/ml NBT and 0.175 μg/ml BCIP in darkness at room temperature until the desired reaction product intensity is achieved. For *in situ* hybridizations using two probes, proceed directly to Step 17 in Method 3.

    If embryos are to be sectioned, we have found that it is best to 'over-develop' the apparent level of signal on whole embryos in order to get sufficient intensity on sections.

15. For *in situ* hybridizations using a single probe, stop the reaction by washing thoroughly in MABT. Refix in 4% paraformaldehyde/1× PBS for 1 h prior to storing or sectioning.

4. Hybridization to thick sections

One limitation of performing hybridizations to whole embryos or tissues is that severe background problems are encountered when using the technique on larger specimens (e.g. embryonic day 13 mice). One way of overcoming this is to cut thick sections (200 μm) of such specimens and perform hybridizations on them. This approach may also be useful for large (e.g. adult) tissues.

Method 5. Hybridization to thick sections

Materials

- Agarose (melting temperature >85 °C, gelling temperature <40 °C)
- 4% Paraformaldehyde/1× PBS
- PBT buffer
- 1× PBS/0.1% (v/v) Tween 20.

Procedure

1. Fix samples in 4% paraformaldehyde/1× PBS overnight.
2. Wash several times in PBT.

3. Melt agarose to make a 4% (w/v) solution in PBS. Place in a 70 °C water bath and allow to equilibrate.
4. Remove embryo/tissue from PBT and remove last traces of PBT with tissue paper and/or a glass pipette.
5. Quickly pour sufficient agarose to cover the specimen in an embedding mould of appropriate size. Immediately submerge specimen in this and orient (if required for sectioning) with a needle or a pair of forceps.
6. Quickly place the mould on ice, taking care that the specimen does not change position. It should take a few minutes for the agarose to set, during which time the agarose around the specimen should be gently swirled occasionally with a pipette to mix in any residual PBT.
7. When set, trim and cut 200 μm sections from the specimen, collecting them in PBT.
8. Proceed to dehydration through a methanol series following steps 4–11 of Method 2.
9. Hybridize and wash using the high stringency protocol described in Method 4.

5. Combined *in situ* hybridization and immunohistochemistry on whole embryos

To compare both mRNA and protein expression patterns, it is possible to follow *in situ* hybridization with immunohistochemistry. Some antibodies are not able to detect the antigen following *in situ* hybridization, and therefore the suitability of individual antibodies should be tested separately. In addition, whole-mount immunohistochemistry for each antibody should be optimized before combining with *in situ* hybridization. We find it easiest to visualize the antibody using a fluorescently conjugated secondary antibody, although using a secondary antibody coupled to horseradish peroxidase allows detection by diaminobenzidine under a brightfield microscope.

The following modifications to the *in situ* hybridization method may increase the subsequent performance of the antibody.

- Fixation
  For *in situ* hybridization, both MEMFA (see Method 2) 4% paraformaldehyde/1× PBS have worked well for us, but fixation should be tested and optimized for each antibody.
- Proteinase K
  The antigen may be destroyed by proteinase K and so a shorter incubation time (5 min) in protease or an alternative permeablization method may improve the signal. We routinely substitute the proteinase K with two 20 min washes in a detergent mix [1% IGEPAL, 1% SDS, 0.5% deoxycholate, 50 mM Tris/HCl (pH 8.0), 1 mM EDTA, 150 mM NaCl].
- Hybridization
  Lowering the hybridization temperature to 60 °C and reducing the prehybridization time to 30 min and the hybridization time to 12 h may increase the preservation of the antigen.
- Detection of mRNA probe
  For single-colour *in situ* protocols it is best to use an alkaline phosphatase-coupled antibody, detected with NBT/BCIP. For double *in situ* hybridization where Fast Red TR/Naphthol AS-MX is used as a stain, it should be noted that this substrate is not permanent and the intensity of colour will fade because of subsequent washing steps in immunohistochemical methods. The following protocol is designed to reduce this by using less detergent. For some antibodies, however, this may increase non-specific binding of antibody. Fast Red is a preferred substrate when FITC-conjugated secondary antibodies are used in the following immunohistochemistry, as it fluoresces at a wavelength of 568 nm, allowing mRNA and protein to be detected at different wavelengths by fluorescence microscopy.

1. Following *in situ* hybridization (according to Methods 2 and 1 or Methods 2 and 4, modified as indicated above), re-fix embryos for 20 min in either MEMFA or 4% paraformaldehyde/1× PBS.

2. Wash the embryos twice in PBT for 5 min.
3. Incubate in PBT/20% sheep serum for 1–2 h at room temperature.
4. Incubate the embryos in the appropriate dilution of primary antibody in PBT/10% sheep serum overnight (or for up to 4 days) at 4 °C with rocking.
5. Wash the embryos six times for 30 min each in PBT/5% (v/v) pre-spun sheep serum.
6. Incubate the embryos in the appropriate dilution of secondary antibody in PBT/10% sheep serum overnight at 4 °C with rocking.
7. Wash the embryos six times for 30 min each in PBT.
8. Equilibrate in 1,4-diazobicyclo-(2,2,2)-octane/glycerol and mount (see Chapter 12 for composition).

6. Sectioning material following *in situ* hybridization

The first procedure described in this section is a rapid method of producing thick sections. The second protocol is a method for producing thin sections as permanent mounts.

Method 6. Cutting thick sections after *in situ* hybridization

Materials

- PBS
- 4% Paraformaldehyde/PBS
- 10% (w/v) Gelatin (Sigma-Aldrich)
- Glycergel (DAKO)
- Nail varnish

Procedure

1. Postfix embryos in 4% paraformaldehyde/PBS for a period of between 1 h and overnight.
2. Wash twice in PBS for 30 min each.
3. Equilibrate the embryos in 10% gelatin in PBS at 37 °C for 1 h.
4. Transfer embryos into a suitable mould for embedding by pouring, and allow the gelatin to set.
5. Carefully cut out the embryo and place the block in 4% paraformaldehyde/PBS to fix the gelatin overnight.
6. Rinse the block in PBS, trim and cut sections to a thickness of 50 μm on a vibrotome.
7. Mount sections on slides in glycerol or glycergel and seal coverslips with nail varnish. Store at 4 °C.

Method 7. Cutting thin sections after *in situ* hybridization

Materials

- 100% Methanol
- Propan-2-ol (Merck BDH)
- Tetrahydronaphthalene (Sigma-Aldrich)

> ***Caution*** *This reagent is very toxic and should only be used in a fume cupboard.*

- Wax (e.g. Paramat pastillated wax, Merck BDH)
- Depex mounting medium (Merck BDH)
- 3-Aminopropyltriethoxysilane (TESPA)-treated microscope slides
- Xylene
- 4% Paraformaldehyde (see Method 2)
- PBS (see Method 2)

Preparing TESPA-treated slides

Microscope slides need to be washed and silanized prior to collection of sections.

1. Place metal racks of 25 slides in 400 ml of 70% (v/v) ethanol/10% (v/v) concentrated HCl for 10 s, then in distilled water for 10 s, and finally in 95% (v/v) ethanol for 10 s.
2. Wrap the rack of slides in aluminium foil to keep them free from dust, and place the slides at 150 °C for 30 min.
3. While still wrapped in foil, cool the slides on the bench.
4. Place the rack of slides in acetone containing 2% (v/v) TESPA for 10 min, then in two changes of acetone for 10 min each, and finally in distilled water for 10 min. This step should be carried out in a fume cupboard.
5. Dry the slides at 37 °C overnight while wrapped in foil. Slides are stored individually in racks (to avoid scratches) at room temperature for up to 3 months.

   The times indicated are for smaller embryos and should be increased for larger material.

   Alternatively, pre-treated slides (e.g. Superfrost, Merck BDH) are now available commercially.

Procedure

1. Postfix embryos in 4% paraformaldehyde/PBS for a minimum of 1 h up to an overnight incubation.
2. Wash the embryos twice in PBS for 15 min.
3. Place the embryos in glass vials and dehydrate in 100% methanol for 5 min.
4. Equilibrate in propan-2-ol for 10 min.
5. Equilibrate in tetrahydronaphthalene twice for 30 min (embryos can be stored here indefinitely).

6. Equilibrate in three changes of wax for 1 h each at 58 °C with occasional shaking.
7. Transfer the embryos to a suitable mould for embedding, orientate using warmed forceps and allow the wax to set. At this point embryos can be stored indefinitely at 4 °C.
8. Cut sections as ribbons at a thickness of 7–20 μm.
9. Collect sections on TESPA-treated slides. This can be done by laying slides on a slide warmer at 50 °C, covering with a pool of distilled water, and floating ribbon lengths on this until dry and creases have disappeared. Allow to dry at 37 °C overnight. Store desiccated at 4 ° C.
10. Dewax in xylene for 2–5 min (prolonged dewaxing results in loss of signal) and mount immediately under Depex.

We thank our colleagues whose suggestions and discussion contributed to the development and revision of these protocols. In particular, Jennifer Walshe introduced the 'orange' colour reaction technology to our laboratory and Monica Ensini provided the protocol for hybridization to thick sections. Our work has been supported by the Medical Research Council (C.I. and I.M.), the Wellcome Trust, the Biotechnology and Biological Sciences Research Council and the Human Frontier Science Program (I.M.).

References

1. Wilkinson, D.G. (1992) in *In Situ* Hybridization. A Practical Approach (Wilkinson, D.G., ed.), IRL Press, Oxford

2. Henrique, D., Adam, J., Myat, A., Chitnis, A., Lewis, J. and Ish-Horowicz, D. (1995) Nature (London) **375**, 787–790

3. Liang, J.O., Etheridge, A., Hantsoo, L., Rubinstein, A.L., Nowak, S.J., Belmonte, J.C.I. and Halpern, M.E. (2000) Development **127**, 5101–5112

Index

Suppliers

Amersham plc
Amersham Place
Little Chalfont
Buckinghamshire HP7 9NA
U.K.
Tel.: +44-1494-544-000
fax: +44-1494-542-266
URL: www.amersham.co.uk

Bachem AG
Hauptstrasse 144
CH-4416 Bubendorf
Switzerland
Tel.: +41-61-935-2323
Fax: +41-61-935-2325
E-mail: sales.ch@bachem.com
URL: www.bachem.com

Bayer plc
Diagnostics Division
Bayer House
Strawberry Hill
Newbury
Berkshire RG14 1JA
U.K.
Tel.: +44-1635-563-000
Fax: +44-1635-563-393
URL: www.bayerdiag.com

Bright Instrument Co. Ltd
St Margarets Way
Huntingdon
Cambridgeshire PE29 6EU
U.K.
Tel.: +44-1480-454528 or 451980 or 451499
Fax: +44-1480-456-031
E-mail:
sales@brightinstruments.com
URL: www.brightinstruments.com

Bunton Instrument Co. Inc.
9607 Doctor Perry Road
Suite 99
Ijamsville
D 21754
U.S.A.
Tel.: +1-301-831-1233
Fax: +1-301-831-1234

CellPath plc
Unit 66
Mochdre Enterprise Park
Newtown
Powys SY16 4LE
U.K
Tel.: 0870-607-0111 (UK)
+44-1686-611-333 (overseas)
Fax: +44-1686-622-946
E-mail: info@cellpath.co.uk

Chroma Technology Corp.
74 Cotton Mill Hill
Brattleboro
VT 05301
U.S.A.
URL: www.chroma.com

3-D Images Ltd
31 The Chine
Grange Park
London N21 2EA
U.K.
Tel. +44-20-8364-0022
Fax: +44-20-8364-1828
E-mail: burder3D@aol.com

3Dx
4850 River Green Parkway
Duluth
GA 30096
U.S.A.
URL: www.3Dx.com

DAKO Corporation
6392 Via Real
Carpinteria
CA 93013
U.S.A.
Tel. +1-805-566-6655
Fax: +1-805-566-6688
E-mail: general@dakousa.com
URL: www.dakousa.com

DAKO Ltd.
Denmark House
Angel Drove
Ely
Cambridgeshire CB7 4ET
U.K.
Tel.: +44-1353-669-911
Fax: +44-1353-668-989
URL: www.dako.co.uk

Dow Corning Ltd
Meriden Business Park
Copse Drive
Allesley
Coventry CV5 9RG
U.K.
Tel.:+44-1676-528-000
Fax:+44-1676-528-001
URL: www.dowcorning.com

Dupont NEN
NEN Life Science Products
BRU/BRU/40349
PO Box 66
Hounslow TW5 9RT
U.K.

Edge 3D Microscopes
520 Washington Boulevard
Marina del Rey
CA 9029
U.S.A.
Tel.: +1-310-578-5175
Fax: +1-240-465-0512

Edmund Scientific Company
101 E. Gloucester Pike
Barrington
NJ 08007-1380
U.S.A.
Tel.: +1-609-573-6250
Fax: +1-609-573-6295

Electron Microscopy Sciences (EMS)
321 Morris Road
Box 251
Fort Washington
PA 19034
U.S.A.
Fax: +1-215-646-8931

Elsa Inc.
2231 Calle De Luna
Santa Clara
CA 95054
U.S.A.
Tel.: +1-408-919-9100
Fax: +1-408-919-9120
URL: www.elsa.com

Enzyme System Products
486 Lindbergh Avenue
Livermoore
CA 94550
U.S.A.
Tel.: 888-44-2664
(toll free; U.S.A. only)
Tel.: +1-925-449-2664
Fax: +1-925-449-1866
URL: www.enzymesys.com

Fisher Scientific UK
Bishop Meadow Road
Loughborough
Leicestershire LE11 5RG
U.K.
Tel.: +44-1509-231-166
Fax: +44-1509-231-893
E-mail: sales@fisher.co.uk
URL: www. sales@fisher.co.uk

Fluka
Sigma-Aldrich Chemical Ltd.
Fancy Road
Poole
Dorset BH12 4QH
U.K.
Tel.: +44-1202-733-114
Fax: +44-1202-715-460
URL: www.sigmaaldrich.com

Gibco—BRL
Life Technologies
3 Fountain Drive
Inchinnan Business Park
Paisley PA4 9RF
U.K.

ICN Biomedicals Inc.
Biomedical Research Products
3300 Hyland Avenue
Costa Mesa
CA 92626
U.S.A.
Fax: 1800-334-6999

ICN Biomedicals Ltd
Unit 18
Thame Park Business Centre
Wenman Road
Thame
Oxfordshire OX9 3XA
U.K.
Fax: +44-1844-263-617

Ilford Imaging UK Ltd
Town Lane
Mobberley
Knutsford
Cheshire WA16 7JL
U.K.

Leica Microsystems (UK) Ltd
Davy Avenue
Knowlhill
Milton Keynes MK5 8LB
U.K.
Tel.: +44-1908-246-246
Fax: +44-1908-609-992

Linscott's Directory
4877 Grange Road
Santa Rosa
CA 95404
U.S.A.
Fax: +1-415-389-6025

Linscott's Directory
Box 188
Falun 79124
Sweden
Fax: +46-23-39232

LIP (Equipment & Services) Ltd
Dockfield Road
Shipley
West Yorkshire BD17 7SJ
U.K.

Meadowlark Optics
P.O. Box 1000
5964 Iris Parkway
Frederick
CO 80530
U.S.A. Tel: +1-303-833-4333
Fax: +1-303-833-4335
E-mail: sales@meadowlark.com
URL: www.meadowlark.com

Merck BDH
Merck House
Poole
Dorset BH15 1TD
U.K.
Tel.: +44-1202-669700
Fax: +44-1202-665-599
URL: www.merckeurolab.ltd.uk (U.K.) or www.bdh.com (non-U.K.)

Molecular Probes Europe BV
PoortGebouw
Rijnsburgerweg 10
2333 AA Leiden
The Netherlands
Tel.: +31-71-523-3378
Fax: +31-71-5233-419
URL: www.probes.com

Molecular Probes Inc.
4849 Pitchford Avenue
Eugene
OR 97402-9165
U.S.A.
Tel.: +1-541-465-8300
Fax: +1-541-344-6504
URL: www.probes.com

National Diagnostics (UK) Ltd
Unit 4
Fleet Business Park
Itlings Lane
Hessle
Hull HU13 9LX
U.K
Tel.: +44-1482-646-020
Fax: +44-1482-646-013
URL:
www.nationaldiagnostics.com

Omega Optical Inc
210 Main Street
Brattleboro
VT 05301
U.S.A.
URL: www.omegafilters.com

Polysciences Inc.
400 Valley Road
Warrington
PA 18976
U.S.A.
Fax: +1-215-343-3291

Promega
Delta House
Chilworth Science Park
Southampton SO16 7NS
U.K.
Tel.: +44-23-8076-0225
Fax: +44-23-8076-7014
URL: www.promega.com

Qiagen Ltd.
Boundary Court
Gatwick Road
Crawley
West Sussex RH10 2AX
U.K.
Tel.: +44-1293-422-911
Fax: +44-1293-422-922
URL: www.qiagen.com

Qiagen Inc.
28159 Avenue Stanford
Valencia
CA 91355
U.S.A.
Tel.: +1-800-426-8157
Fax: +1-800-718-2056
URL: www.qiagen.com

R&D Systems Europe Ltd
19 Barton Lane
Abingdon Science Park
Abingdon
Oxfordshire OX14 3NB
U.K.
Tel.: +44-1235-529-449
Fax: +44-1235-533-420
E-mail: info@rndsystems.co.uk
URL: www.rndsystems.com

R&D Systems
614 McKinley Place NE
Minneapolis
MN 55413
U.S.A.
Tel.: +1-800-343-7475
or +1-612-379-2956
Fax: +1-612-379-6580
URL: www.rndsystems.com

Raymond A. Lamb Ltd
Units 4 and 5
Parkview Industrial Estate
Lottbridge Drove
Eastbourne
Sussex BN23 6QE
U.K.
Tel.: +44-1323-737-000
Fax: +44-1323-733-000
E-mail: sales@ralamb.com
URL: www.ralamb.net

Real 3D Enterprises Inc.
P.O. Box 2368
Culver City
CA 90231
U.S.A.

Roche Products Ltd
40 Broadwater Road
Welwyn Garden City
Hertfordshire AL7 3AY
U.K
Tel.: +44-1707-366-000
fax: +44-1707-338-297
URL: www.roche.com

Schott
Werk Weisbaden-Faseroptik
Postfach 130367
6200 Weisbaden 13
Germany

Shandon Lipshaw
Customer Service Department
171 Industry Drive
Pittsburgh
PA 15275
U.S.A.
Fax: +1-412-788-1138

Shandon
Life Sciences International (UK)
Unit 5
The Ringway Centre
Edison Road
Basingstoke
Hampshire RG21 6YH
U.K.
Fax: +44-1256-817-292

Sigma-Aldrich Chemical Ltd
Fancy Road
Poole
Dorset BH12 4QH
U.K.
Tel.: +44-1202-733-114
Fax: +44-1202-715-460
URL: www.sigmaaldrich.com

Sigma-Aldrich Corp.
P.O. Box 14508
St Louis
MO 63178
U.S.A.
Tel.: +1-314-771-5765
Fax: +1-314-771-5757
URL: www.sigmaaldrich.com

StereoGraphics Corporation
2171 E. Francisco Boulevard
San Rafael
CA 94901
U.S.A.
Tel.: +1-415-459-4500
Fax: +1-415-459-3020
URL: www.stereographics.com

Stratagene
11011 N. Torrey Pines Road
La Jolla
CA 92037
U.S.A.
Tel.: +1-800-424-5444
URL: www.stratagene.com

TAAB Laboratories Equipment
3 Minerva House
Calleva Road
Aldermaston
Berkshire RG7 8NA
U.K.
Fax: +44-118-981-7811

Thermo Shandon
Chadwick Road
Runcorn
Cheshire WA7 1PR
U.K.
Tel.: +44-1928-562-541

U.H.I. Systems K.K.
Room 918
Kioi-cho TBR Building
5–7 Koji-Machi
Chiyoda-ku
Tokyo 102
Japan
Tel.: +81-3-3263-7161
Fax: +81-3-3263-7176
E-mail: microscope@uhi.co.jp

VayTek Inc.
305 West Lowe Avenue
Fairfield
IA
U.S.A.
Tel.: +1-515-472-2227
Fax: +1-515-472-8131

Vector Laboratories Inc.
30 Ingold Road
Burlingame
CA 94010
U.S.A.
Fax: +1-650-697-0339

Vector Laboratories Ltd
3 Accent Park
Bakewell Road
Orton Southgate
Peterborough PE2 6XS
U.K.
Fax: +44-1733-237-119

VWR Canlab
2360 Argentia Road
Mississauga
ON
Canada L5N 5Z7
Fax: +1-905-821-3460